Practical Scientific Computing

Related titles:

Essentials of scientific computing: numerical methods for science and engineering
(ISBN 978-1-904275-32-9)
Modern development of science and technology is based to a large degree on computer modelling. To understand the principles and techniques of computer modelling, students should first get a strong background in classical numerical methods, which are the subject of this book. This text is intended for use in a numerical methods course for engineering and science students, but will also be useful as a handbook on numerical techniques for research students. *Essentials of scientific computing* is as self-contained as possible and considers a variety of methods for each type of problem discussed. It covers the basic ideas of numerical techniques, including iterative process, extrapolation and matrix factorization, and practical implementation of the methods shown is explained through numerous examples. An introduction to MATLAB is included, together with a brief overview of modern software widely used in scientific computations

Fundamentals of university mathematics
Third edition
(ISBN 978-0-85709-223-6)
The third edition of this popular and effective textbook provides in one volume a unified treatment of topics essential for first-year university students studying for degrees in mathematics. Students of computer science, physics and statistics will also find this book a helpful guide to all the basic mathematics they require. It clearly and comprehensively covers much of the material that other textbooks tend to assume, assisting students in the transition to university-level mathematics. An essential reference for first-year university students in mathematics and related disciplines, it will also be of interest to professionals seeking a useful guide to mathematics at this level and capable pre-university students.

Mathematical analysis and proof
Second edition
(ISBN 978-1-904275-40-4)
This fundamental and straightforward text addresses a weakness observed among present-day students, namely a lack of familiarity with formal proof. Beginning with the idea of mathematical proof and the need for it, associated technical and logical skills are developed with care and then brought to bear on the core material of analysis in such a lucid presentation that the development reads naturally and in a straightforward progression. Retaining the core text, the second edition has additional worked examples which users have indicated a need for, in addition to more emphasis on how analysis can be used to tell the accuracy of the approximations to the quantities of interest which arise in analytical limits.

Details of these and other Woodhead Publishing mathematics books can be obtained by:

- visiting our web site at www.woodheadpublishing.com
- contacting Customer Services (e-mail: sales@woodheadpublishing.com; fax: +44 (0) 1223 832819; tel.: +44 (0) 1223 499140 ext. 130; address: Woodhead Publishing Limited, 80 High Street, Sawston, Cambridge CB22 3HJ, UK)

If you would like to receive information on forthcoming titles, please send your address details to: Francis Dodds (address, tel. and fax as above; e-mail: francis.dodds@woodheadpublishing.com). Please confirm which subject areas you are interested in.

Practical Scientific Computing

Ali Muhammad and Victor Zalizniak

WP
WOODHEAD
PUBLISHING

Oxford Cambridge Philadelphia New Delhi

Published by Woodhead Publishing Limited,
80 High Street, Sawston, Cambridge, CB22 3HJ, UK
www.woodheadpublishing.com

Woodhead Publishing, 1518 Walnut Street, Suite 1100, Philadelphia,
PA 19102-3406, USA

Woodhead Publishing India Private Limited, G-2, Vardaan House, 7/28 Ansari Road,
Daryaganj, New Delhi – 110002, India
www.woodheadpublishingindia.com

First published 2011, Woodhead Publishing Limited

British Library Cataloguing in Publication Data
A catalogue record for this book is available from the British Library.

ISBN 978-0-85709-225-0 (print)
ISBN 978-0-85709-226-7 (online)

The publisher's policy is to use permanent paper from mills that operate a sustainable forestry policy, and which has been manufactured from pulp which is processed using acid-free and elemental chlorine-free practices. Furthermore, the publisher ensures that the text paper and cover board used have met acceptable environmental accreditation standards.

Table of Contents

Preface

Scientific computing is about developing mathematical models, numerical methods and computer implementations to study and solve real problems in science, engineering, business and even social sciences. Mathematical modeling requires deep understanding of classical numerical methods. There are a number of commercial and open source tools available which provide a rich collection of built-in mathematical functions. These tools are widely used by researchers to model, solve and simulate mathematical problems. These tools also offer easy to use programming languages for users to develop more functions.

This book is divided into two parts. In the first part, we present an open source tool, numEclipse. It is a Java based tool modeled after MATLAB® and it is implemented as a plug-in for Eclipse, which is a leading integrated development environment (IDE) for Java programming. In the second part, we study the classical methods of numerical analysis. We present the numerical algorithms and their implementations using numEclipse.

Although we have tried to make this book as self-contained as possible, knowledge of calculus, linear algebra and differential equations is a prerequisite. Some exposure to a programming language will be helpful to follow the computer programs.

Such a small volume as this book cannot do justice to the vast area of classical numerical methods. The choice of some of the topics is based on our own preferences and teaching experience. This book provides enough foundations for a beginner to venture into more advanced texts in the subject area.

Ali Muhammad
Victor Zalizniak

Acknowledgements

Ali would like to thank his wife Samira, son Danish and daughter Amna for their encouragement and patience while he ignored them for days at a time. He would also like to thank his parents Abdul Sattar and Zohra Sattar for their efforts in raising him with the desire to learn and to propagate knowledge.

Victor would like to express deepest gratitude to his parents Ludmila and Evgeniy Zalizniak for their patience and sacrifices in raising him and for their encouragement to learn.

Finally, we would like to thank the team at Horwood and Woodhead Publishing for their efforts in making this book a reality.

Part I

In this part, we introduce an open source scientific tool, numEclipse. Readers already familiar with MATLAB[1] or Octave[2] and who do not wish to learn a new tool could just read the first chapter and skip to the second part of the book. Those who are interested in learning this tool could read through chapters 1 to 3. Readers interested in programming using MATLAB/Octave like scripting language or even high-level programming languages will benefit from chapter 4. Ambitious readers who want to understand the internals of the tool and who are interested in modifying the basic linear algebra operations are invited to read chapter 5. Finally, anybody interested in generating pretty plots must read chapter 6.

---

[1] http://www.mathworks.com
[2] http://www.gnu.org/software/octave

1

Introduction

Computers were initially developed with the intention to solve numerical problems for military applications. The programs were written in absolute numeric machine language and punched in paper tapes. The process of writing programs, scheduling the execution in a batch mode and collection of results was long and complex. This tedious process did not allow any room for making mistakes. The late 1950s saw the beginning of research into numerical linear algebra. The emergence of FORTRAN as a language for scientific computation triggered the development of matrix computation libraries, i.e., EISPACK, LINPACK. The availability of these libraries did not ease the process of writing programs. Programmers still had to go through the cycle of writing, executing, collecting results and debugging. In the late 1970s, Cleve Moler developed the first version of MATLAB to ease the pain of his students. This version did not allow for M-files or toolboxes but it did have a set of 80 useful functions with support for matrix data type. The original version was developed in FORTRAN. The first commercial version released in 1984 was developed in C with support for M-files, toolboxes and plotting functions. MATLAB provided an interactive interface to the EISPACK and LINPACK. This eliminated the development cycle and the users were able to view the results of their commands immediately due to the very nature of an interpreter. Today, it is a mature scientific computing environment with millions of users worldwide.

With all the good things that it offers, MATLAB is accessible to only those users who can afford to purchase the expensive license. There has been a number of attempts to develop an open source clone for MATLAB. The most notable among them are GNU Octave, Scilab and RLab. They provide matrices as a data type, support complex numbers, offer a rich set of mathematical functions and have the ability to define new functions using scripting language similar to MATLAB. There are many other tools in this domain and numEclipse is a new entrant in this arena.

numEclipse is built as a plug-in for eclipse so before we delve into the details of numEclipse we need to look at eclipse. It is generally known as an Integrated Development Environment (IDE) for Java. In fact, it is more than just an IDE; it is a framework which can be extended to develop almost any application. The framework allows development of IDE for any programming language as a plug-in. The Java development support is also provided through a built-in plug-in called Java Development Toolkit (JDT). Today a number of programming languages like C/C++, Fortran, etc. are supported by eclipse. The aim behind the development of numEclipse as an eclipse plug-in was to develop an IDE for scientific computing. In the world of scientific tools, IDE means interactive development environment rather than integrated development environment. If you look at 3Ms (MATLAB, Mathematica & Maple) they are highly interactive due to the very nature of an interpreter and geared towards computational experimentations by individual users rather than supporting team based project development. The development of a scientific application could be a very complex task involving a large team with the need to support multiple versions of the application. This could not be achieved without a proper IDE with the notion of project and integration to source control repository. Sometimes, it is also desirable to write programs or functions in a high-level programming language other than the native scripting language offered by the tool. A good IDE should provide the ability to write programs in more than one language with the support to compile, link, execute and test the programs within the same IDE. Fortunately, the design decision to develop numEclipse as an eclipse plug-in enabled all these capabilities. numEclipse implements a subset of MATLAB and GNU Octave's scripting language, m-script, this allows development of modules in specialized areas like MATLAB toolboxes. The pluggable architecture provides the ability to override the basic mathematical operations like matrix multiplication and addition.

In the following sections, we will learn to create a numEclipse project. We will look at the user interface including the numEclipse perspective and related views. We will also learn about using the interpreter for interactive computation as we would do in MATLAB or Octave. We will write and execute a program written in m-script which demonstrates the plotting features of numEclipse.

1.1 Getting Started

In this section, we will introduce the numEclipse working environment. We will learn how to create a numEclipse project. We will review the user interface including numEclipse perspective and views. Installation of numEclipse plug-in requires the latest java runtime environment and Graphical Modeling Framework (GMF) based eclipse installation. A step by step guide for installation and configuration is provided on the project website (http://www.numeclipse.org). Once the plug-in is successfully installed, configured and verified, we are ready to create our first project. This two step process is described as follows. Select eclipse menu *File → New → Project*. It will bring up *New Project* dialog box. Select the wizard *numEclipse Project* under the category *numEclipse* and click on *Next* button, as shown in figure 1.1. It will bring up the *Project Wizard* for numEclipse. Type the project name in the textbox and click on *Finish* button, as shown in figure 1.2. On successful completion of the above steps, you will see that a project has been created in your workspace. Also, you will notice that the Perspective is changed from Java to numEclipse, as shown in figure 1.3.

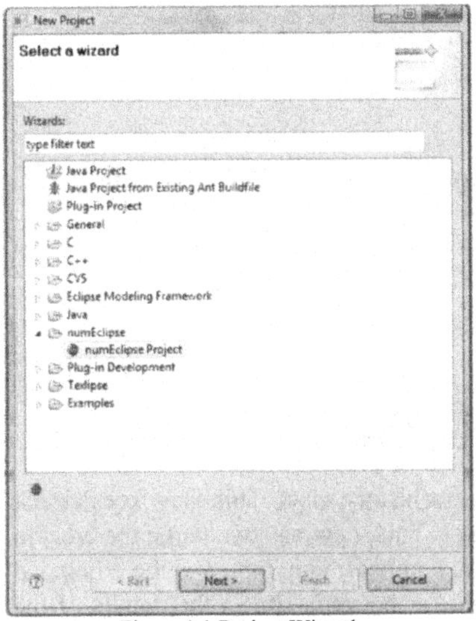

Figure 1.1 Project Wizard.

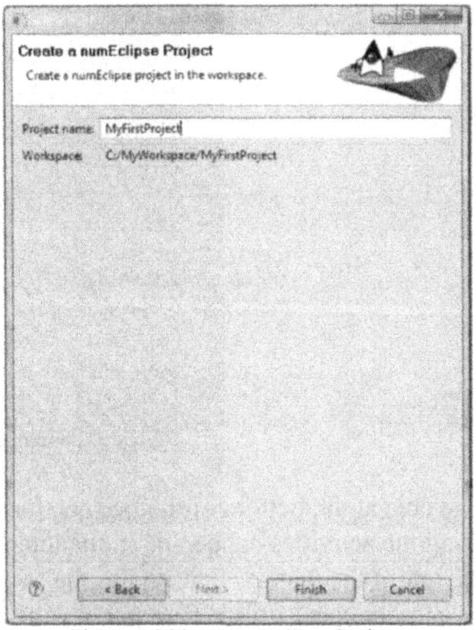

Figure 1.2 New numEclipse Project.

Before looking at the numEclipse perspective (figure 1.3), it is important to understand the project structure just created. The navigator view on the left of the perspective shows the project. It consists of two folders (i.e. "Interpreter" and "Source") and a default interpreter *default.i* created under folder *Interpreter*. If the navigator view's filter does not block *.resources* files then you will also be able to see the *.project* file created for the numEclipse project. The central area of the window shows the *default.i* interpreter. This is where most of the user interaction happens. Here you will type the commands and look at the results, as shown in the following listing (listing 1.1).

Listing 1.1

```
>> x = [1 2; 3 4];
>> y = x ^ 2
y =
7  10
15  22
```

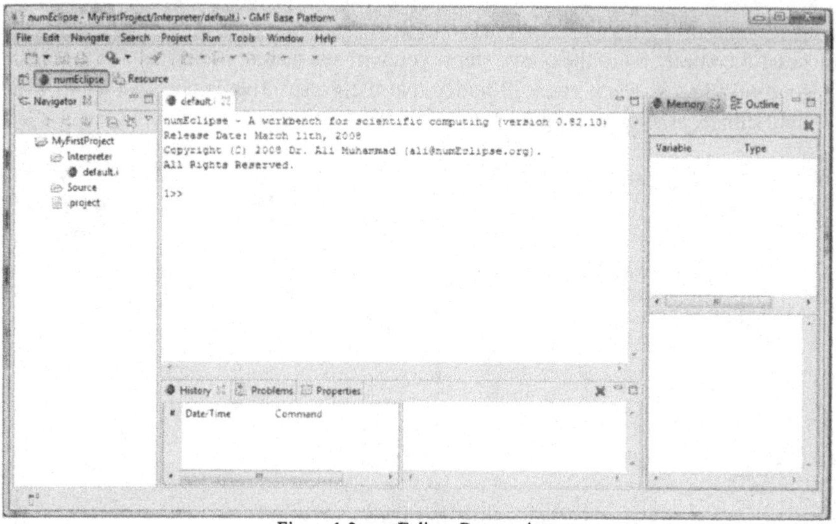

Figure 1.3 numEclipse Perspective.

The area at the bottom of the perspective shows history view. This view keeps track of all the activities happening in the interpreter. This view has two parts; the area on the left of the history view shows the sequence number, date/time and the command in chronological order. The area on the right of the view shows the details of the command selected on the left, as shown in figure 1.4.

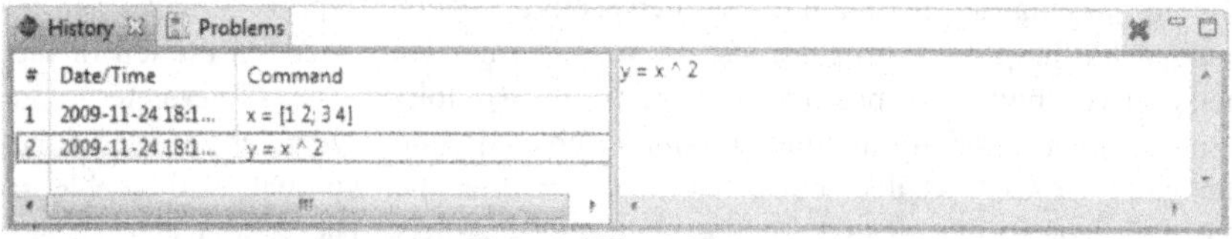

Figure 1.4 History View.

The cross × button on the top right of the view is used to clear the history view. It will erase all the entries in the view. A user can select the contents of the window on the right and copy it to the interpreter to run the command. The area on the right of the perspective is the memory view. This view shows all the variables in the memory of the interpreter. This view also has two parts. The part on the top shows the list of variables and their corresponding types. When a variable is selected in this part of the view, the corresponding value of that variable is shown in the bottom part of the view. Like the history view, the cross × button on the top right of the memory view is used to clear the memory. It will erase all the entries in the view.

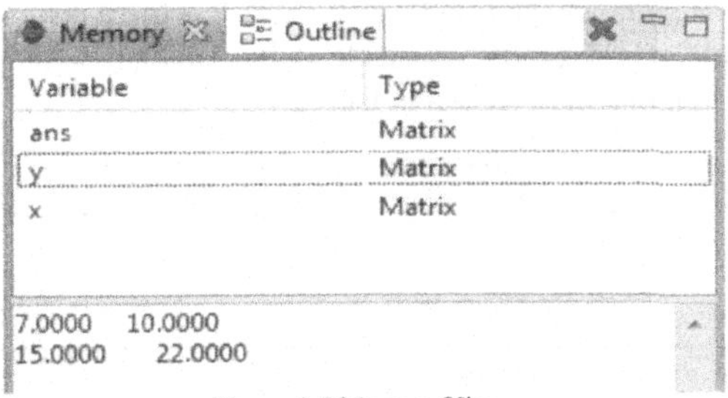

Figure 1.5 Memory View.

It might seem like a lot of work for somebody who is new to scientific computing or more familiar with MATLAB. But, it must be remembered that this tool offers more than an interpreter and it is more geared towards scientific application development. There are a lot of benefits of project oriented approach even for an end user. Unlike MATLAB, you can open more than one interpreter at a time. Say you wanted to work on calculus, right click on Interpreter folder and create a simple file *calculus.i*. This will open another interpreter window in the center of the perspective. A project allows you to open as many interpreter windows you need. In MATLAB, you will have to launch a new instance of the application in order to get another interpreter window. The memory and history views are associated to the active interpreter. When you switch the interpreter window, the memory and history views will reflect the activity in the current active window. You can also save the session for each interpreter using menu *File* → *Save*. Next time when you launch the application, it will load all the variables and history from the saved session file.

1.2 Interpreter

numEclipse can be used in two modes; interactive and programming. In the interactive mode, we type the commands in the interpreter window and execute them as we go from one command to another. This mode is useful when you are experimenting with your algorithms or you are using the tool as a sophisticated scientific calculator. In the programming mode, you would write a set of commands / statements in a file and execute all the commands in the file together. The commands / statements in both modes are almost same, as they are based on the same language specification. In interactive mode, you have some additional commands specific for the interactive environment which is not applicable to the programming environment. The interpreter allows you to write and test mathematical expressions and small blocks of code. One can start typing the commands directly after the prompt >> as shown in listing 1.2.

Listing 1.2

```
>> x = 0:10
x =
0.00001.00002.00003.00004.00005.00006.00007.00008.00009.000010.0000
```

It creates an array *x* with values from *0* to *10*. Note that there is no need to declare or initialize a variable. The type of a variable is determined by the use and it could change with next command, as shown in listing 1.3.

Listing 1.3

```
>> x = 'Hello World'
x =
Hello World
```

Now, the type has changed to string and the same variable *x* has a value *Hello World*. Variable name should always start with a letter and it is case-sensitive. Each line that we type in the interpreter is captured by the history view and each new variable introduced in the interpreter is shown in the memory view. This could be used for the debugging purposes.

Listing 1.4

```
>> t = 0:0.1:pi;
```

In the example shown in listing 1.4, an array *t* is created with values ranging from *0* to *pi* with the increment of *0.1*. Here *pi* is a mathematical constant. You would notice that the output of the expression is not echoed back to the screen. This is due to the semicolon at the end of the line. numEclipse has some built-in variables and constants as given in table 1.1.

Table 1.1 Special variables and constants

Name	Description	Value
realmin	Smallest positive number	4.9000E-324
realmax	Biggest positive number	1.7977E308
eps	Precision of a number	2.2204E-016
ans	Last result	
Inf or –Inf	Positive or negative infinity	
NaN	Not a number	
i or j	Iota	$\sqrt{-1}$

numEclipse has a very interesting feature which is not available in MATLAB/Octave. It allows you to define your own constants. Select eclipse menu *Windows* → *Preferences*, it will bring up the *Preferences* dialog box. On the left of the dialog box, expand the category *numEclipse* → *Preferences* and select *Constants*, as shown in figure 1.6. Here you can add or modify constants and they immediately become available in all the interpreter instances.

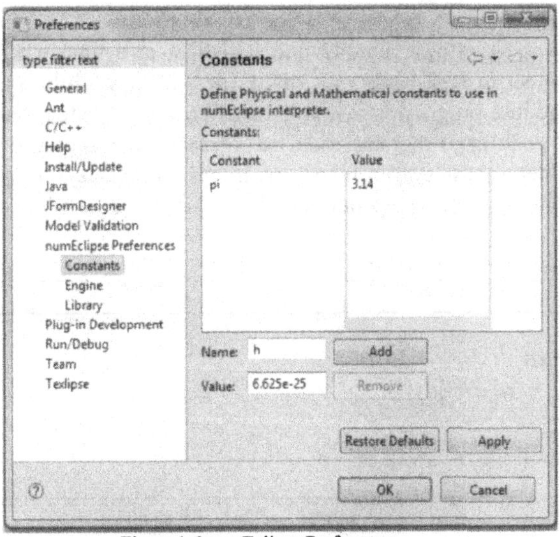

Figure 1.6 numEclipse Preferences.

So it is important as part of configuration to define the constants required for your specific work.

Listing 1.5

```
>> y = sin(t);
>> plot(t, y);
```

Expression in listing 1.5 creates a new array *y* which contains the *sin* values of the previously defined array *t*. numEclipse provides integration with gnuplot. The steps to configure link to gnuplot are described on project website. *plot* command opens a gnuplot window if it is not already initialized and it draws a line plot of variable *y* versus corresponding values of *t*, as shown in following figure 1.7. Chapter 7 covers the plotting functions in detail.

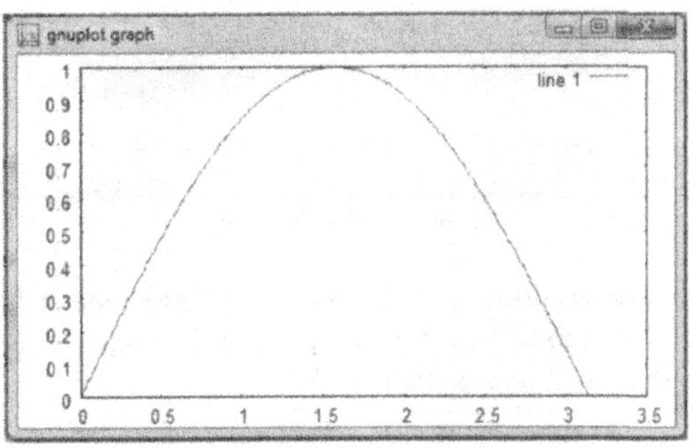

Figure 1.7 *y = sin(t)*.

1.3 Program

The interpreter allows running one statement at a time whereas a program contains a number of statements in a single file and runs all the statements at one time. A numEclipse program is a simple text file with *.m* extension. It must reside under the *Source* folder. Programs could be organized by creating sub-folders within the *Source* folder to any level of hierarchy. In order to create a program, right click on the source folder and create a new simple file *firstprog.m*. Add the code in listing 1.6 to the file.

Listing 1.6

```
for i = 1 : 10
   for j = 1 : 10
      h (i, j) = 1 / (i + j + 1);
   end;
end;
plot(h);
```

Once it is saved, it can be used from any instance of the interpreter. Either open up the default interpreter or create a new interpreter as described earlier. Enter the command in listing 1.7 to the command prompt.

Listing 1.7

```
>> run firstprog.m
```

The GNU Plot window is shown the following figure 1.8.

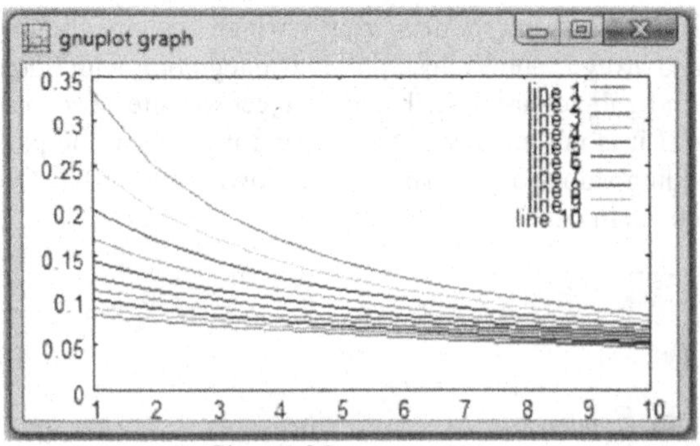

Figure 1.8 Program output.

You can use either the memory view on the right-hand side of the perspective to examine the value of variable *h* used in the program or you can just enter *h* on the prompt to get the value as shown in listing 1.8.

Listing 1.8

```
>> h
0.3333 0.2500 0.2000 0.1667 0.1429 0.1250 0.1111 ...
```

Note that the program does not have any structure like *begin* to indicate the start of the program or *end* to indicate the end of the program. It is just a collection of commands that you could also execute from the interpreter.

2

Expressions

In the previous chapter, we learnt about the fundamentals of numEclipse workbench. We looked at the numEclipse project, perspective, interpreter and how to write and execute a simple program in m-script. This chapter and the next chapter focus on the programming language. A programming language has two major parts, either it is evaluating or it is execution. The evaluation part is related with the expressions. For example, $y = x + 2$ is an expression where the program evaluates the value of y by using the value of x. The execution part of the program is related with statements. For example, *write ("Hello World")* is a statement which prints the string "Hello World!" on the screen.

In this chapter, we will focus on different types of expressions. An expression involves variables of different type and operators. So we will look at the data types and operators applicable to those data types.

2.1 Matrix

It is the key data type of this language. There is no equivalent for this data type in standard programming languages. A matrix with only one column represents a mathematical entity known as "column vector" and a matrix with one row represents a "row vector". A matrix is basically a two dimensional array of elements. These elements could be real numbers, complex numbers, strings, booleans, structures or cells. Multi-dimensional matrices or arrays are not supported at this point.

Listing 2.1

```
>> A = [1 2 3; 4 5 6; 7 8 9]
A =
1.0000  2.0000  3.0000
```

```
4.0000 5.0000 6.0000
7.0000 8.0000 9.0000

>> B = [-9 7 4
2 -6 4
1 3 9]
B =
-9.0000 7.0000 4.0000
2.0000 -6.0000 4.0000
1.0000 3.0000 9.0000
```

In the above listing, for the first matrix *A* we use semicolons to separate rows whereas in matrix *B* we just use a new line to indicate the start of a new row. In the above example, we used spaces to separate the elements in a row. You can also use *comma* to avoid any ambiguity. A single real number, complex number and even a boolean is treated as matrix of size 1×1.

Every variable in numEclipse is a matrix. The following example shows how to modify and access an element of a matrix.

Listing 2.2

```
>> A(2, 2) = 3.14
A =
1.0000 2.0000 3.0000
4.0000 3.1400 6.0000
7.0000 8.0000 9.0000

>> x = A(1, 3)
x =
3.0000
>> A([2 3], [1 2])
ans =
4.0000 3.1400
7.0000 8.0000
```

An individual element of a matrix is referenced by providing the row and column index in parentheses as shown in the above example. Also note in the last example that we can also reference a sub-matrix by providing an array of indices.

2.2 Real Number

Real number is a double precision floating point number. It is equivalent to java primitive "double". The maximum value of a real number is defined by a constant *realmax*, *1.7977e+308*, the minimum value is defined by another constant *realmin*, *4.9000e-324* and the precision of the real number is defined by *eps*, *2.2204e-016*. The following listing shows some examples of expressions with real numbers.

Listing 2.3

```
>> x = 3.5
x =
3.5000

>> y = 2.7
y =
2.7000

>> z = x + y
z =
6.2000

>> realmin
realmin =
4.9000E-324

>> realmax
realmax =
1.7977E308

>> eps
eps =
2.2204E-016
```

2.3 Complex Number

A complex number has no equivalent in standard programming languages. It is simply a pair of real numbers. The first part is called the real part and the second part is called the imaginary part. The following listing shows some examples of complex numbers.

Listing 2.4

```
>> z1 = 2 + 2j
z1 = 2.0000 + 2.0000i

>> z2 = -1 + 3i
z2 = -1.0000 + 3.0000i

>> z3 = z1 * z2
z3 =
-8.0000 + 4.0000i
```

You would notice in the above example that we use both *"i"* and *"j"* to represent the imaginary part. In the following listing, we show how to get the real and imaginary part of a complex number.

Listing 2.5

```
>> z = 2 – i
z =
2.0000 - 1.0000i

>> x = real(z)
x =
2.0000

>> y = imag(z)
y =
-1.0000
```

numEclipse supports a number of other functions to support complex number computation, a few of them are listed in the following table 2.1.

Table 2.1 Complex functions

Function	Description
conj	Returns the complex conjugate
real	Returns the real part
imag	Returns the imaginary part
abs	Returns the magnitude of the complex number
angle	Returns the angle of the polar representation of the complex number

2.4 Boolean

The boolean data type corresponds to the java boolean primitive type. A boolean type variable is limited to only *true* and *false* values. An expression evaluating into a number greater than zero will be treated as a true value and a zero value expression will be treated as false. In the following, we give a simple example of relational expression which evaluates to true value.

Listing 2.6

```
>> b = 2 > 1
b =
1.0000
>> [1 2] < 0
ans =
0.0000   0.0000
```

In the second example in the above listing, when we compare the array with zero value a component wise comparison is performed and *0s* are returned to represent false values. We can always use the literals *true* and *false* along with *0*s and *1*s in a boolean expression. It is demonstrated in the following example.

Listing 2.7

```
>> b = 2 > 1 & false
b =
0.0000
>> [1 1] & true
ans =
1.0000   1.0000
>> [0 1] | false
ans =
0.0000   1.0000
```

Boolean values, operators and expressions play an important role in loops and branching statements, we will shortly learn more about them.

2.5 String

The String data type is equivalent to java String. numEclipse allows both single and double quotes to enclose string literals. Unlike MATLAB, a numEclipse string is not an array of characters, it is an object. You can create an array of strings with numEclipse but it is very different from the array of characters in MATLAB. In the following, we create two string variables and we show how they could be concatenated.

Listing 2.8

```
>> s1 = 'John'
s1 =
John
>> s2 = 'Hello'
s2 =
Hello
>> s = s2 + ' ' + s1
s =
Hello John
```

numEclipse offers a number of useful string processing functions. In the following example, we show how to compare two strings.

Listing 2.9

```
>> strcmp('John', 'john')
ans =
false
>> strcmpi('John', 'john')
```

```
ans =
true
```

The two functions return different truth values because *strcmpi* is case insensitive.

Listing 2.10

```
>> index('12345678901234567890', '456')
ans =
4.0000
>> rindex('12345678901234567890', '456')
ans =
14.0000
```

In the above example, *index* returns the index of first appearance of second string within first string whereas *rindex* returns the index of last occurrence.

2.6 Structure

Structure is a composite data type. It is very similar to *struct* in C language. Unlike a matrix, a structure can contain data elements of various types. Elements / fields of a structure are named rather than indexed. The individual fields are addressed using the dot notation as shown below.

<structure_name>.<field_name>

A structure within an array of structures must have the same set of fields. A single structure is treated as an array of size 1×1. The following example shows how to create a structure.

Listing 2.11

```
>> employee.name = 'John';
>> employee.id = 123;
>> employee.salary = 3000;
>> employee.pref = [10 20; 30 40];
>> employee
employee =
perf :
10.0000  20.0000
30.0000  40.0000
name : John
id : 123.0000
salary : 3000.0000
```

In the above example, we use the assignment to create structure fields. There is no need to declare the structure. The following example shows the use of struct function to create a structure variable.

Listing 2.12

```
>> emp = struct('name', 'John', 'id', 123, 'salary',
3000)
emp =
name : John
id : 123.0000
salary : 3000.0000
```

This function takes an even number of arguments as a sequence of field name and value pairs.

struct('name1', value1, 'name2', value2, ...)

Note that the name of a field is provided as a string. The following example shows how to access the value of a structure field.

Listing 2.13

```
>> emp.name = 'John';
>> emp.id = 123;
>> emp.salary = 3000;
>> emp.pref = [10 20; 30 40];
>> x = emp.salary
x =
3000.0000
>> s = emp.name
s =
John
```

Similarly, accessing the value of a field from a structure array is shown below.

Listing 2.14

```
>> emp(2).name = 'Peter';
>> emp(2)id = 456;
>> emp(2).salary = 2500;
>> emp(2).perf = [1 2; 3 4];
>> name = emp(2).name
name =
Peter
>> sal = emp(1).salary
sal =
3000.0000
>> x = emp(2).perf(2,2)
x =
4.0000
```

In the above example, we also show how to get a matrix element from a matrix field value of a structure array. Unlike Matlab, there are no add or remove functions available in numEclipse. You can add a field as shown in above example. There is no method available to remove a field. Using the above example, we look at the *size* function in the following.

Listing 2.15

```
>> size(emp)
ans =
1.0000   2.0000
>> size(emp.pref)
ans =
2.0000   2.0000
```

The first example above returns the size of the structure array *emp* whereas the second example returns the size of the matrix field *pref* within the *emp* structure. numEclipse does not supported nested structures at this point.

2.7 Cell

Cell array is another composite data type. It is very similar to structure array except the way it organizes the data. Unlike a structure array, each cell in an array could have an entirely different type of element. A cell array can be created using a function as shown in the following example.

Listing 2.16

```
>> c = cell(2,2);
>> c(1, 1) = {'Hello World'};
>> c(1, 2) = {2};
>> c(2, 1) = {[1 2]};
>> c(2, 2) = {1 + 2i};
>> c
c =
Cell(1, 1)
Hello World
Cell(1, 2)
2.0000
Cell(2, 1)
1.0000   2.0000
Cell(2, 2)
1.0000 + 2.0000i
```

There is no need to use a function; a cell can be created just by using assignment statement. In the following, we show how to create a cell using curly brackets.

Listing 2.17

```
>> ce = {[3.13   2.71]}
```

```
ce =
3.1300  2.7100
```

Obtaining data from a cell element is shown in the following example.

Listing 2.18

```
>> x = c{2, 2}
x =
1.0000 + 2.0000i
```

A cell array cannot be nested in another cell but it can contain structure array. Here is another example based on the cell variable *"c"* defined earlier.

Listing 2.19

```
>> c(1,2) = struct('name', 'John', 'id', 123, 'salary', 3000)
c =
Cell(1, 1)
Hello World
C(1, 2)
name : John
id : 123.0000
salary : 3000
Cell(2, 1)
1.0000  2.0000  3.0000
Cell(2, 2)
1.0000 + 2.0000i
```

The above examples make it clear that cell arrays are very flexible and they can hold any amount of data of any type. This feature makes them candidates for input and output variables for functions. This will be discussed in detail in the later chapters.

2.8 Range Expression

A range expression provides a short cut notation to define a vector using colon ":" operator. Say we want to define a vector containing positive integers ranging from *1* to *10*, using this notation we could define it as follows.

Listing 2.20

```
>> m = 1:10
m =
1.0000 2.0000 3.0000 4.0000 5.0000 6.0000 7.0000
8.0000 9.0000 10.0000
```

We could also define a vector from *1* to *50* with increments of *10* using two colon operators such that the number between the colons defines the increment value. It is demonstrated in the following listing.

Listing 2.21

```
>> n = 1:10:50
n =
1.0000 11.0000 21.0000 31.0000 41.0000
```

This saves us from listing all the vector elements. Similarly, we could define vectors with descending values using negative incremental steps.

Listing 2.22

```
>> u = 10:-2:1
u =
10.0000 8.0000 6.0000 4.0000 2.0000
```

The colon operator could also be used as a wildcard to select rows and columns of a matrix. When we use the colon for a subscript, it represents the entire row or column. For example, *A(3, :)* represents the 3rd row of matrix *A* and *A(: , 2)* represents the 2nd column.

2.9 Boolean Expression

A boolean expression is composed of boolean primitives *"true"* and *"false"* and logical operators. numEclipse supports following boolean operators.

Table 2.2 Boolean operators

Name	Notation
NOT	~
AND	&&, &
OR	||, |

The unary NOT operator negates the truth value of a variable. The binary AND operator returns true if and only if both operands are true. The binary OR operator returns false if and only if both operands are false. Both AND and OR operators are short circuit operators which means they will not evaluate the second operand if the first is enough to evaluate the resultant. In the following, we show some examples of boolean expressions.

Listing 2.23

```
>> true && false
ans =
false
```

```
>> true | false
ans =
true
>> ~true
ans =
false
>> ~false
ans =
true
```

2.10 Relational Expression

A relational expression is composed of numbers (scalar, vector, matrix or complex) and relational operators. A relational expression evaluates to a boolean value. numEclipse supports following relational operators.

Table 2.3 Relational operators

Name	Notation
less than	<
greater than	>
less than and equal to	<=
greater than and equal to	>=
equal to	==
not equal to	<>, !=, ~=

In the following, we show some examples of relational expressions.

Listing 2.24

```
>> 3 > 1
ans =
true
>> A = rand(2)
A =
0.7242  0.2868
0.9793  0.9371
>> A > 0.5
ans =
1.0000  0.0000
1.0000  1.0000
```

It is important to note that both boolean and relational expressions evaluate to a boolean value. So they could be combined into more complex expressions as shown in the following example.

Listing 2.25

```
>> x = rand(2)
x =
0.82740.0648
0.45710.3825
>> y = rand(2)
y =
0.9761     0.4578
0.7462     0.6924
>> x > 0.5 && y > 0.5
ans =
1.0000   0.0000
0.0000   0.0000
```

2.11 Numerical Expression

A numerical expression involves variables, literal numbers and arithmetic operators. A variable or number could be a real number, complex number, vector or a matrix. numEclipse supports the arithmetic operators given in table 2.4. The operators addition, subtraction and power are obvious. numEclipse provides two types of divisions, left and right division. Left division divides the right argument and right division divides the left argument. Component-wise operators are applied on the elements of the operands. This could be explained by an example, as shown below.

Table 2.4 Arithmetic operators.

Name	**Notation**
addition	+
subtraction	-
multiplication	*
right division	/
left division	\
power	^
component-wise multiplication	.*
component-wise right division	./
component-wise left division	.\
component-wise power	.^

Here, when we apply dot power on matrix *A*, it does not square the matrix rather it squares each element of the matrix. Similar, when we dot multiply *A* and *B*, it is not a matrix multiplication rather corresponding elements of the matrices are multiplied.

Listing 2.26

```
>> A = [1, 2; 3, 4];
>> A.^2
```

```
ans =
1.0000 4.0000
9.0000 16.0000
>> B = eye(2)
B =
1.0000 0.0000
0.0000 1.0000
>> A.*B
ans =
1.0000 0.0000
0.0000 4.0000
```

In the following, we present a rather complicated numerical expression.

Listing 2.27

```
>> A
A =
0.7242 0.2868
0.9793 0.9373
>> z
z =
1.0000 - i
>> b = (A + 1) /2 - z ^ 2
b =
0.8621 + 2.0000i 0.6434 + 2.0000i
0.9896 + 2.0000i 0.9687 + 2.0000i
```

3

Statements

In the previous chapter, we learnt about the data types, expressions and operators supported by numEclipse. In this chapter, we will focus on statements including assignment statement, loops and conditional statements.

3.1 Assignment Statement

The assignment statement is the most simple statement in a programming language. In numEclipse an assignment could take different forms depending on the type of variables involved. In its simple form, it looks like the following expression.

<var_name>=<expression>

"=" is the assignment operator. This statement evaluates the expression and assigns the resulting value to the variable named on the left hand side. The left hand side or lvalue must be a valid identifier name. Identifier naming must start with a letter and it could be followed by zero or more letters, numbers or underscores in any combination. There is no need to declare a variable before assigning a value. numEclipse automatically determines the data type from the right hand side expression (or rvalue). The type of a variable could change as we assign a new value of different type to an existing variable. In the following, we show a few examples of assignments.

Listing 3.1

```
>> X = 1.0
X =
1.0000
>> Y = [1 2 3; 4 5 6]
Y =
1.0000  2.0000  3.0000
```

```
4.0000  5.0000  6.0000
>> S = 'Hello World'
S =
Hello World
>> B = 1 < 2
B =
1.0000
```

If a simple assignment is ended with a semi-colon then the interpreter will not report the value of the right hand side expression on the screen. A matrix assignment involves assigning an element of an existing matrix to a value. In case of a column or row vector, the assignment will be as follows.

<var_name> (<expression>) = <expression>

The expression within parentheses on the left hand side is an index value so it should evaluate to an integer value. In the case of an out of bound index value, the variable will be resized to accommodate the index value. The index value starts from "*1*". In the case of a two dimensional matrix, the assignment will look like the following.

<var_name>(<expression>, <expression>) = <expression>

The first index in the parentheses determines the row number and the second index determines the column number of the matrix element. Here is an example for matrix element assignment.

Listing 3.2

```
>> A = rand(3)
A =
0.8467 0.0952 0.2836
0.9565 0.3782 0.5011
0.7280 0.2052 0.6668
>> A(2, 3) = 1.5
A =
0.8467 0.0952 0.2836
0.9565 0.3782 1.5000
0.7280 0.2052 0.6668
```

numEclipse functions could return one or more variables so the syntax for assigning multiple return values is shown in the following.

[<var_1> <var_2> … <var_n>] = <expression>

The expression on the right-hand side will always be a function call. In the following, we provide an example of multiple returned variables.

Listing 3.3

```
>> x = min([2 1 3]])
```

```
x =
1.0000
>> [x idx] = min([2 1 3]])
x =
1.0000
idx =
2.0000
```

In the first assignment example above, function *min* returns only the minimum value of the matrix element. In the second example, it returns the minimum value as well as the index of the minimum value. Not only a numEclipse function can return multiple values, it can also determine from the function call how many variables need to be returned on a particular call.

3.2 Loop Statements

Loop statements allow a program to run a block of statements a fixed number of times. numEclipse supports two loop statements, *for* loop and *while* loop.

3.2.1 "for" Statement

"for loop" is a compound statement. It is a repetitive statement since it allows a group of statements to be executed a number of times. The structure of this statement type is as follows.

```
for <identifier> = <range expression>
      <statement>
      <statement>
          .
          .
      <statement>
end/endfor
```

A *"for"* statement should always be closed with either *"end"* or *"endfor"* keyword. The *"identifier"* works like a loop counter. The range expression evaluates to vector which determines the values taken by the identifier. Here is an example of a *"for"* loop statement.

Listing 3.4

```
>> x = 0;
>> for i = 1 : 10
   x = x + 1;
end;
>> x
x =
10
```

In the above example variable *x* is assigned a value *"0"*. The *"for"* loop is repeated *10* times and at each iteration *x* is incremented by *1*.

3.2.2 "while" Statement

"while loop" is also a compound statement and it is repetitive in nature. The difference between a *"for"* and *"while"* loop is that the *for* loop executes statements for a fixed number of time whereas the *while* loop executes all grouped statements as long as a condition is satisfied. The structure of a while statement is as follows.

```
while <conditional expression>
      <statement>
      <statement>
          .
          .
      <statement>
end/endwhile
```

A *"while"* must be closed with either an *"end"* or *"endwhile"* keyword. In this statement the conditional expression is evaluated on the beginning of each iteration, if the conditional expression evaluates to true then the enclosed statements are executed otherwise the loop ends. Here is an example of a *while* loop.

Listing 3.5

```
>> y = [0 0 0];
>> i = 1;
>> x = 3;
>> while x > 0
   y(i) = x;
   i = i + 1;
   x = x - 1;
end;
>> x
x =
0.0000
>> y
y =
3.0000 2.0000 1.0000
```

In the above example, three statements are executed as long as the condition $x > 0$ is satisfied.

3.3 Conditional Statements

Conditional statements execute when a condition is satisfied. numEclipse supports two conditional statements, *"if"* and *"switch"* statements.

3.3.1 "if" Statement

An *"if"* statement is used to implement decisions in a program. It consists of one or more conditional expressions and corresponding block of statements. If a conditional expression evaluates to true then the corresponding block of statements is executed otherwise the next conditional expression is evaluated, if present.

```
if <conditional expression>
      <statement>
          .
          .
      <statement>
elseif <conditional expression>
      <statement>
          .
          .
      <statement>
  .
  .
else
      <statement>
          .
          .
      <statement>
end/endif
```

In the above structure *"if"*, *"elseif"*, *"else"*, *"end"* and *"endif"* are the keywords. There could be zero or more *"elseif"* sections in an *"if"* statement. There could be zero or one *"else"* section in an *"if"* statement. Blocks of statements associated with *"else"* are only executed when none of the other conditions are satisfied.

An *"if"* statement could be nested within another *"if"* statement. It could increase the complexity but sometimes it is desirable. Remember that an inner *"if"* statement is evaluated before an outer statement.

Listing 3.6

```
x = input('Enter a number');
if (x>0)
  sign = 1;
else if (x<0)
  sign = -1;
else
  sign = 0;
```

In the above program, the user is asked to enter a number and it is assigned to variable *x*. If *x* is greater than *0* the variable sign is assigned the value *1*; in the case where *x* is less than *0* the variable sign is set to *-1*, otherwise it is set to *0*.

3.3.2 "switch" Statement

A *"switch"* statement is also used to implement decisions in a program. It is helpful to use *"switch"* when you have a lot of *"elseif"* statements within an *"if"* statement. It switches among several cases based on an expression, as shown in the following syntax.

```
switch <switch expression>
    case <case expression>
        <statement>
        .
        .
        <statement>
    case <case expression>
        <statement>
        .
        .
        <statement>
    .
    .
    otherwise
        <statement>
        .
        .
        <statement>
end/endswitch
```

The *"switch"* expression should be either a string or a scalar value. The *"case"* expression should evaluate to the same data type as *"switch"* expression. When a *"case"* expression matches the *"switch"* condition, the corresponding block of statements is executed. Unlike C or Java, you do not need to use *break* statement only first matching case is executed. Here is an example.

Listing 3.7

```
money = input('Enter an amount (in cents)');
switch money
case 1
  cents = cents + 1;
case 5
  nickel = nickel + 1;
case 10
  dime = dime + 1;
case 25
  quarter = quarter + 1;
case 100
  dollars = dollars + 1;
```

```
otherwise
  alert('invalid amount');
end;
```

3.4 Continue and Break Statements

In the absence of *"goto"* statements, *"continue"* and *"break"* statements are sometimes very useful. The continue statement moves the control to the next step of the *"for"* or *"while"* loop. For example,

Listing 3.8

```
>> for i = 1 : 100
sum = sum + 1;
end;
>> sum
sum =
100.0000
>> sum = 0;
>> for i = 1 : 100
   if rem(i, 2) == 0
     continue;
   end;
   sum = sum + 1;
  end;
>> sum
sum =
50.0000
```

The above example shows how the continue statement skips an iteration for an even number of values of *"i"* using condition *rem(i, 2) == 0*.
The break statement ends the execution of a *"for"* or *"while"* loop as shown below.

Listing 3.9

```
>> sum = 0;
>> for i = 1 : 100
    sum = sum + 1;
    if i == 50
      break;
    endif;
  endfor;
>> sum
sum =
50
```

The above example shows how the break statement ends the execution of the *"for"* loop when the value of *"i"* becomes *50*.

4

Programming

In the last two chapters, we learnt the specification of numEclipse's scripting language, m-script. In this chapter, we will learn how to put the script in action to write a program, function and procedure. Most of the programming goals could be achieved using m-script. Sometimes we might want to develop a function in higher level programming language to achieve better performance or to integrate with an existing set of APIs. numEclipse is developed using java so it provides seamless integration with java functions. A java program can access C functions using Java Native Interface JNI. numEclipse takes advantage of this mechanism and provides the ability to integrate C functions. In this chapter, we will also learn how to develop and deploy Java and C functions.

4.1 Program

It was previously mentioned that numEclipse could be used in both interactive and programming mode. We saw a number of examples of how the interpreter could be used interactively. We also looked at an example of m-script program in chapter 1. Unlike other programming languages, an m-script program does not have strict structure. It does not have *"begin"* or *"end"* statements or curly brackets to show the start and end of the program. The program is stored in a text file with *".m"* extension, i.e., m-file. It contains a sequence of statements which you could also execute individually from the command prompt. In that sense, it is very much like a batch file. It also makes the program much simpler. A programmer will normally try different commands interactively. Once, he is sure about the sequence of commands to achieve the objectives, he could copy the commands from the *"history view"* to an m-file within the Source folder of the project. The program could be executed from the command prompt using the *"run"* command. A script program runs in the same workspace as the interpreter. It means that all the existing variables in the memory will be available to the program. Similarly, all the variables created during the execution of the program will be available in workspace memory even after the completion of program execution. This could also cause potential conflicts among variables with the same name. To avoid such a situation, we could use the "clear" command in the beginning of the program. This command removes all variables from the workspace memory. In the following, we present a simple program which generates Fibonacci numbers based on user input. Fibonacci numbers are generated based on a recurrence relation as defined in the following.

$$\begin{aligned} f_1 &= 1 \\ f_2 &= 1 \\ f_{n+2} &= f_n + f_{n+1} \end{aligned} \tag{4.1}$$

Listing 4.1 (fibonacci.m)

```
1  n = input('Number of Fibonacci numbers');
2  N = str2num(n);
3  fib = zeros(N, 1);
4  fib(1) = 1;
5  fib(2) = 1;
6  for i=3:N
7      fib(i) = fib(i-2) + fib(i-1);
8  end;
9  ?fib
```

In the above program, line 1 asks the user to provide the number of Fibonacci numbers to generate. Line 2, converts the user input from string to a number. Lines 3 and 4, initialize the first two values of Fibonacci numbers as defined in the recursive formula given above. Lines 6–8, use a *"for"* loop to generate the numbers. The last line, 9, displays the values of vector "fib". Note the question mark in the last line; it is used to display the value of the variable. In the interactive mode, you do not need to use a question mark to display the value but in a program it is required.

Listing 4.2

```
run Fibonacci.m
```

The program becomes immediately available for use in any interpreter in the project. The above code shows how to invoke the program. All the programs must be saved in the *"Source"* folder within the project. Unlike MATLAB, these programs cannot be compiled into MEX files. Although the compiled program could speed up the execution of a program, it does not make sense to compile code for an interpreter. If you really need performance then you should consider writing the program in Java or C as described later in the chapter.

4.2 Function

The ability to create user-defined functions is available in all programming languages including m-script. Most of the programming in numEclipse involves writing functions. A *"toolbox"* is a collection of functions in a particular subject area. For example, an image processing toolbox will contain functions related with image processing. A lot of built-in toolbox functions within numEclipse are also implemented using m-script.

In general, a function takes zero or more arguments as input and returns one output value. Here you can go beyond this and define functions which could take a variable number of arguments and return a variable number of output values. In the following, we will show the specification and example of each possible case. Here is a simple specification of a function.

```
function identifier = name (arg_1, arg_2, ..., arg_n)
    <statement>
    <statement>
        .
        .
        .
    <statement>
end/endfunction
```

In a simple case, a function takes a fixed number of input arguments and returns a single output value. The following example presents a simple function.

Listing 4.2

```
function y = sqr(x)
y = x.^2;
```

You would notice that unlike other programming languages there is no need to use a *"return"* statement. The output variable is directly assigned the value in the function body. The *"return"* statement is available but it is used differently from other programming languages. It is used to stop the execution of the function or procedure. The program execution is returned to the calling program, when a *"return"* statement is encountered in a program. Each function call creates a separate workspace memory for the function. The actual input arguments are copied to this workspace. It means that a pass-by-value approach is used instead of pass-by-reference. The advantage is that the function could alter the input parameters within the function but it will not affect the parameter values in the calling program or function. Say, for some reason you want to make a variable available to the interpreter's memory space, it is possible by declaring the variable as global variable within the function. For example,

```
global planck_constant;
planck_constant = 6.626e-34;
```

Like a program, a function is also created in the m-file. The file name must be same as the function name for the sake of traceability. The interpreter uses the file name as the function name. In MATLAB or Octave, you could define multiple functions or procedures in the same file; numEclipse does not allow this for the sake of simplicity. Like a program, the function file must reside under the *"Source"* folder, within any nested folder if required. It also becomes immediately available for use in any interpreter within the same project. Here is an example to test the function.

Listing 4.3

```
>> sqr(2)
ans =
4.0000
>> z = 1 + 2i;
>> sqr(z)
ans =
-3.0000 + 4.0000i
```

In the above test example, note that the same function works for both a real number and complex number argument. It will even work for a matrix, as mentioned before, every number in numEclipse is treated as a matrix.

The second syntax definition in the following shows that a function can return more than one value in an array.

```
function [id_1, id_2, ..., id_n] = name (arg_1, arg_2, ..., arg_n)
        <statement>
        <statement>
            .
            .
            .
        <statement>
end/endfunction
```

The following code shows the example of a function with multiple returned values. Note that there is no *"end"* or *"endfunction"* to close the function. Unlike statements, use of *"end"* is optional in functions and procedures.

Listing 4.4

```
function [K, T, B, C] = KTBC(n)
% Create the four special matrices assuming n > 1
K = toeplitz([2 -1 zeros(1, n-2)]);
T = K; T(1, 1) = 1;
B = K; B(1, 1) = 1; B(n, n) = 1;
C = K; C(1, n) = -1; C(n, 1) = -1;
```

The comment on the second line of the function definition is used as a help document in MATLAB and Octave. numEclipse has not implemented a help system yet. But for the sake of documentation, it is a good idea to put some comments describing the function.

The following listing shows how to call these functions.

Listing 4.5

```
>> x = 1:10;

>> var = variance(x)
var =
2.8723

>> [k t b c] = KTBC(2)
k =
2.0000  -1.0000
-1.0000  2.0000

t =
1.0000  -1.0000
-1.0000  2.0000

b =
1.0000  -1.0000
-1.0000  1.0000

c =
2.0000  -1.0000
-1.0000  2.0000
```

Most mathematical functions have zero or more input arguments and a single output argument. A user of these functions will always pass the fixed number of arguments and expect single output. Here is another example,

Listing 4.6

```
% min.m
function z = min(x, y)
% This function returns the minimum of x or y.
if (x <= y)
   z = x;
else
   z = y;
end;
```

This function accepts two input arguments, *x* and *y*, and it returns the minimum of these two arguments. The *"end"* keyword in above function is the end of the *"if"* statement rather than the function. In fact, the *"end"* keyword for the function is optional. You can also use *"endfunction"* like in octave script. One most important point, there is no mention of input arguments' data type. This is really wonderful, this function will work for real numbers, complex numbers, row/column vectors and matrices. The relational operator "<=" automatically overload based on the type of operands.

In the following we show some examples of the function call.

Listing 4.7

```
>> min(2, 3)
ans =
2.0000

>> min([1 2], [3 4])
ans =
1.0000   2.0000
```

"min" function as described above does not put any constraints on the type of the input arguments. This might be a problem in certain situations for example when somebody tries to find a minimum of a string and a matrix. In that case, you will use the *"error"* function as shown below.

Listing 4.8

```
% min.m
function z = min(x, y)
% This function returns the minimum of x or y.
if type(x) ~= type(y)
   error('usage min(x, y): both x and y must be same type.');
end;
if (x <= y)
  z = x;
else
  z = y;
end;
```

In this modified definition, we call an error function when the type of input arguments does not match. The *"error"* function returns the control back to the caller program along with its string argument. Another function, *"warning"*, could also be used in the situation where you want to send a string message back to the caller program without returning the control, so the function will keep executing.

The above function restricts the number of input arguments to two. Say, we want to modify it in such a way that it could accept any number of inputs and return the minimum of those input arguments. This is how we will do it.

Listing 4.9

```
% min.m
function z = min(varargin)
% This function returns the minimum of input arguments.
z = varargin{1};
for i = 2 : nargin
   if (varargin{i} <= z)
      z = varargin{i};
   end;
end;
```

Listing 4.10

```
>> m = min(1, 2, 3, 4, 5)
m =
1.0000
```

When the above function is called as shown above, all the input arguments are packed into a cell array *"varargin"* and an automatic variable *"nargin"* is created with value *5* which reflects the number of inputs. The ith argument could be accessed using the curly bracket, *varargin{i}*.

Now, we want to modify this function to return more than one value. Say, we want it to return both the minimum and maximum values.

Listing 4.11

```
function [min max] = extreme(varargin)
% This function returns the minimum and maximum of input arguments.
min = varargin{1};
max = varargin{1};
for i = 2 : nargin
   if (varargin{i} < min)
      min = varargin{i};
   elseif (varargin{i} > max)
      max = varargin{i};
   end;
end;
```

This function could be called as shown below.

Listing 4.12

```
>> [m n] = extreme(1, 2, 3, 4, 5)
m =
1.0000
n =
5.0000
```

If we call the same function slightly differently, as follows.

Listing 4.13

```
>> [m n p] = extreme(1, 2, 3, 4, 5)
```

We will get an array index out of range error. This situation could be avoided by using the *"nargout"* as shown below.

Listing 4.14

```
function [min max] = extreme(varargin)
% This function returns the minimum and maximum of input arguments.
if nargout > 2
  error('usage [min, max] =  extreme(varargin): wrong number of output
     arguments');
end;
min = varargin{1};
max = varargin{1};
for i = 2 : nargin
   if (varargin{i} < min)
      min = varargin{i};
   elseif (varargin{i} > max)
      max = varargin{i};
   end;
end;
```

The *"nargout"* contains the number of output variables requested by the caller program. Checking this variable enables the function to printout a meaningful error message. Let's give another twist to this function.

Listing 4.15

```
function m = extreme(s, varargin)
% This function returns the minimum and maximum of input
% arguments.
min = varargin{1};
max = varargin{1};
n = length(varargin);
```

```
for i = 2 : n
    if (varargin{i} < min)
        min = varargin{i};
    elseif (varargin{i} > max)
        max = varargin{i};
    end;
end;
if s == 'min'
  m = min;
elseif s == 'max'
  m = max;
else
  error('usage extreme(s, varargin): s should be either
       max or min.');
end;
```

And here is the call to this function.

Listing 4.16

```
>> extreme('min', 1, 2, 3, 4, 5)
 ans =
1.0000

>> extreme('max', 1, 2, 3, 4, 5)
ans =
5.0000
```

In this function, the first input argument is a string which determines if the function should return the minimum or maximum of the rest of the input arguments. This example is presented to show that *"varargin"* could be used with other input arguments. Similarly, *"varargout"* could be used with other output arguments. But, we should always use these at the end of the list of the arguments, for obvious reason. Another important point to remember is that in this situation *"nargin"* and *"nargout"* will return the total number of input and output arguments and it will not be the same as the length of these cell arrays. So if you want to loop around these arrays then it is better to use the *"length"* function on these cell arrays rather than the automatic variables, *"nargin"* and *"nargout"*.

A function could also be referenced using a function handle. This allows us to call the original function indirectly through its handle. The function handle is created using the "@" symbol in front of the function name.

<function handle> = @<function name>

e.g. *f = @sin*

The advantage of this feature is that we can create a function of function by passing a function handle to another function. In the following function, we calculate the first difference values for the given input function.

Listing 4.17

```
function y = difference(f)
% This function returns the first difference vector of the input vector.
len = length(f);
y(1:len) = 0;
for i=1:len-1
   y(i) = f(i+1) – f(i);
end;
```

In the following, we demonstrate how two different functions are passed as input arguments to the above function. This feature enables us to write generalized mathematical functions which will work with any input function, for example, functions to integrate, differentiate, or function to solve differential equations.

Listing 4.18

```
>> f = @sin;
>> difference(f)
ans =
0.0678  -0.7682  -0.8979  -0.2021  0.6795 0.9364  0.3324  -0.5772
-0.9561  -0.4560
>> g = @sqrt;
>> difference(g)
ans =
0.4142  0.3178  0.2679  0.2361  0.2134
0.1963  0.1827  0.1716  0.1623  0.1543
```

Once all the user defined functions are developed and tested, you might want to deploy them as a toolbox for other users. As mentioned earlier, numEclipse does not support compilation of m-script functions to MEX files. But, it does have a neat mechanism of deployment. The deployment package is just a zip file containing the user-defined functions. So you compress the m-files containing the functions in a zip file and deliver to the end-users.

The end-user should take the following steps to include these functions into the interpreter.

1. Copy the zip file locally to the hard-disk.
2. Open the eclipse preferences using menu *Windows → Preferences*.
3. Select the numEclipse preference category and then select the item Library.
4. Click on button "*New*" and select the zip file.

5. You will be prompted to re-start the application. You do not have to re-start if you are adding completely new functions. If you were modifying an existing function then you must re-start.

To be on the safe side it is always recommended to re-start, if you are not sure.

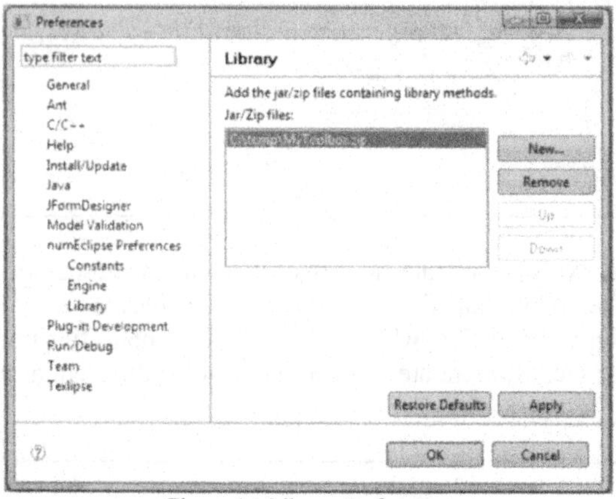

Figure 4.1 Library Preference.

This completes the deployment. You are ready to try it now. Now we deleted the *"extreme.m"* file from the workspace and deployed the zip file. Then we tried the function again and the result is same as before.

Listing 4.17

```
>> extreme('max', 1, 2, 3, 4, 5)
ans =
5.0000
```

Say, you had two definitions of the same function. One in the workspace and other one deployed as a library. The interpreter will pick the definition from the workspace since it has higher precedence.

4.3 Procedure

A procedure is same as a function except it does not return any value. The syntax definition for a procedure is shown below.

procedure name (arg_1, arg_2, ..., arg_n)
<statement>
<statement>
.
.
.

<statement>
end/endfunction

Here is an example of a procedure.

Listing 4.18

```
procedure proc(x)
  plot(1:length(x), x);
end;
```

In the following, we show how to call a procedure.

Listing 4.19

```
>> proc((1:10).^2)
```

And it results in the following plot.

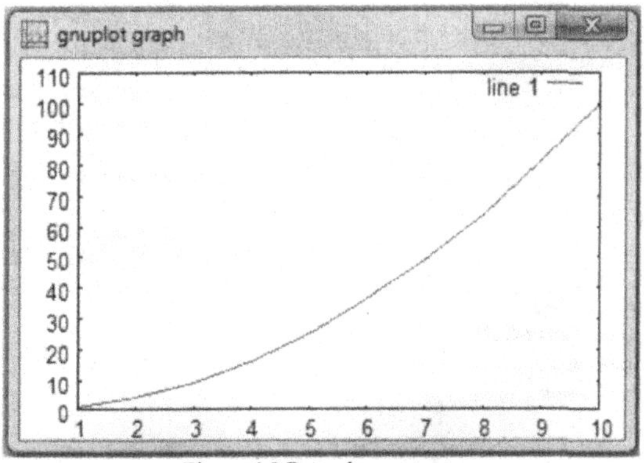

Figure 4.2 Procedure output.

Similar to a function, we could also create a procedure with variable number of input arguments. We also have the automatic *"nargin"* and *"varargin"* available within a procedure. The only difference from a function is that we cannot return a variable. Although it is not good programming practice, we can still make a procedure communicate with another procedure using the global variables. A procedure also starts with the keyword *"function"* and it is also written in an m-file. The packaging and deployment process also remains the same.

4.4 Java Programming

It is assumed that the reader is familiar with the java development using eclipse. To develop java functions in numEclipse, you would also need to understand some internals of this application. Before we go into those details, let us start with a simple example to demonstrate the development process.

Select eclipse menu *File → New Project*, it will bring up *"New Project"* wizard. Select *"Java Project"* and click on *"Next"* button. Type the name of project on the *"New Java Project"* dialog box and click on *"Finish"* button to create the new java project. You would need to follow similar steps to create the package *"test"* and a java class *"MyJavaClass"* within the package. In the following figure, we show a java project *"MyJavaProject"* containing a class *"MyJavaClass"* and a static method *"addOne"*. Note that, in the *"Referenced Libraries"* of project *"MyJavaProject"* you need to add the *"library.jar"* file. It is available within the numEclipse plug-in. You will not be able to compile the class without this library in the classpath. Once the class is successfully compiled, you can immediately start to use it in the interpreter. Right-click in the numEclipse project *"MyFirstProject"* and select *"Properties"* from the pop-up menu. Select *"Project References"* on the *"Properties"* dialog box and select *"MyJavaProject"*. This establishes the link between the two projects. Now, switch back to any interpreter and try out the function *"addOne"* as shown in the following.

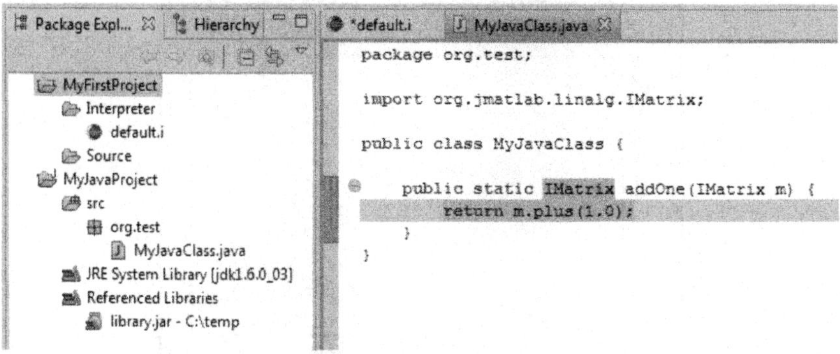

Figure 4.3 MyJavaClass.

Listing 4.20

```
>> addOne(1);
>> ans =
2.0000
>> addOne(1+j)
>> ans =
2.0000 + i
>> addOne([1 2; 3 4])
>> ans =
2.0000  3.0000
4.0000  5.0000
```

Let us look at the function *"addOne"*, it is basically a static function. It takes an argument of type *"IMatrix"*. The function body contains only one line, which calls a method *"plus"* on the input object with the value *"1.0"* and returns the result of type *"IMatrix"*. *"IMatrix"* is an interface for the *"Matrix"* data type discussed in chapter 2. Each data type in numEclipse has a corresponding java interface, e.g., IComplex, IStruct, ICell. These interfaces are defined on the numEclipse project website (http://www.numeclipse.org/interface.html). It is important to know the methods available in each interface but there is no need to understand the actual implementation. numEclipse provides a default implementation of these data types and the pluggable architecture allows replacement of this implementation. It will be discussed later in detail in another chapter. numEclipse also provides the interface class *"LinearAlgebra"*. The implementation of this class provides methods to create variables of different data types. The numEclipse interpreter keeps all the variables in a look up table called *"Symbol Table"*. Each variable in the table is wrapped in another object called *"Symbol"*. This *"Symbol"* class provides methods to set and get the variables. In the case of an exception, a java function should throw *"SemanticException"*. Since it is a *"RuntimeException"* there is no need to use *"try ... catch"* block. The interpreter will catch the exception and it will show the error message on the command line.

In the last example, we defined the function to accept a matrix argument and return a matrix. But when we called the function, we passed double, complex and matrix numbers and it worked for all of these data types. It is important to remember that a double or a complex variable is treated as matrix of dimension 1×1. So you only need to define one function for all three types. In case, you want your function to work with any other types then you should provide the overloaded definition of the function for each of those types.

The *"MyJavaClass"* needs some changes before it could be deployed. The modified class is shown in the following.

Listing 4.21

```
package test;
import org.numEclipse.toolbox.Toolbox;
import org.numEclipse.linalg.IMatrix;
public class MyJavaClass implements Toolbox {
private static LinearAlgebra alg;
    public static void setAlgebra (LinearAlgebra a) {
        alg = a;
    }
    public static IMatrix addOne (IMatrix x) {
        return x.plus(1.0);
    }
}
```

Note that, we added a static variable *"alg"* of type *"LinearAlgebra"*. We also added a static method to set the value of this variable. These additions facilitate the pluggable implementation of linear algebra functions. The interpreter could change the implementation on the fly. The class now implements an interface *"Toolbox"*. Any java class that we want to deploy must implement this interface. In fact, there is not much to implement because the *"Toolbox"* interface is empty. The *"function library manager"* within the interpreter uses a common mechanism of java reflection to load all the functions whether they are built-in or deployed. When a class is implementing the "Toolbox" interface, the interpreter could identify the deployed functions.

Deployment of java functions is very similar to the deployment of m-script. Here, we export the java functions to a jar file using the eclipse *File → Export* menu. Then add the jar file to the numEclipse library using the preferences as we did in the previous section. Make sure that the numEclipse project does not depend on the java project in the workspace. Let the workbench restart and test the functions from the interpreter. In case of duplication, function definition in the workspace takes precedence over the deployed function.

Let us look at some more examples.

Listing 4.22

```
1  public static IMatrix min(IMatrix m1, IMatrix m2) {
2      IMatrix m = m1.lessThan(m2);
3      if (m.isAllPositive()) {
4          return alg.createMatrix(m1);
5      } else {
6          return alg.createMatrix(m2);
7      }
8  }
```

This function returns the minimum of two input arguments. In lines 2, we compare the input matrix arguments. The result of the comparison is a matrix of zeros and ones. *"One"* when the corresponding element is less than the element in m2 and *"zero"* otherwise. In line 3, we check if all elements of the resultant matrix are greater than zero and return the result accordingly. Note that we return a new instance of the input variable rather than the actual variable. This ensures that the original variables are not modified in the symbol table. This function is very similar to the function developed using m-script in the previous chapter. Let's see how the function call works here.

Listing 4.23

```
>> min(2, 3)
ans =
```

```
2.0000
>> min([1 2], [3 4])
ans =
1.0000   2.0000
```

Interestingly, the function works for both double and matrix input arguments. We can use any combination of double, complex and matrix input pair and it will still work. So there is no difference in the way it works regardless of the implementation language. In the following we show how to raise an error or exception.

Listing 4.24

```
public static IMatrix min(IMatrix m1, IMatrix m2) {
  int rows1 = m1.getRows();
  int cols1 = m1.getCols();
  int rows2 = m2.getRows();
  int cols2 = m2.getCols();
  if (rows1 != rows2 || cols1 != cols2)
    throw new SemanticException("input argument size must be the same.");
       IMatrix m = m1.lessThan(m2);
       if (m.isAllPositive()) {
              return alg.createMatrix(m1);
        } else {
              return alg.createMatrix(m2);
        }
}
```

In this modified function, we compare the size of input matrices and throw *"SemanticException"* in case of a mismatch. Here is an example to call this function,

Listing 4.25

```
>> min([1 2], 3)
input argument size must be the same.
```

Note that the error message shown on the command line is same as the string argument passed to the *"SemanticException"* in the function. It is always a good idea to use a meaningful message. Now, we will modify this function in such a way that it will accept a variable number of input arguments. Here is a code listing,

Listing 4.26

```
public static IMatrix min (List list) {
  IMatrix result = ((Symbol)list.get(0)).getMatrix();
  for (int i=1; i<list.size(); i++) {
```

```
    IMatrix m2 = ((Symbol) list.get(i)).getMatrix();
    IMatrix m = m2.lessThan(result);
    if (m.isAllPositive()) {
      result = m2;
    }
  }
  return alg.createMatrix(result);
}
```

In this example, we used a *List* instead of *IMatrix*. Basically, when we call this function from the interpreter or any other function or program, the interpreter packs all the input arguments in a list. We extract the contents of the list to compute the result. Note that, we do not return an object passed to the function. We always create a new object out of the resultant matrix object. This ensures the integrity of the variables in the workspace.

We will change this function further to return multiple output parameters. Here is the code listing,

Listing 4.27

```
public static List extreme(List list) {
  List result = new ArrayList();
  IMatrix min = ((Symbol) list.get(0)).getMatrix();
  IMatrix max = ((Symbol) list.get(0)).getMatrix();
  for (int i=1; i<list.size(); i++) {
    IMatrix m2 = ((Symbol)list.get(i)).getMatrix();
    IMatrix m = m2.lessThan(min);
    IMatrix n = m2.greaterThan(max);
    if (m.isAllPositive()) {
      min = m2;
    } else if (n.isAllPositive()) {
      max = m2;
    }
  }
  result.add (alg.createMatrix(min));
  result.add (alg.createMatrix(max));
  return result;
}
```

This method could take variable number of arguments and it could return more than one output parameters. The input and output parameters are packed into lists by the interpreter as mentioned in the previous example. Here is an example to call this method.

Listing 4.28

```
>> extreme(1, 2, 3, 4, 5)
ans =
1.0000
>> [m n] = extreme(1, 2, 3)
m =
1.0000
n =
3.0000
```

The result is same as we saw in the last chapter. Essentially, you could achieve the same functionality either you write the function in m-script or java. Of course, the java functions will be faster than the m-script.

4.5 C Programming

This section is intended to show you how to write C functions for numEclipse. It is more difficult of the last two approaches but luckily eclipse comes with some tools which make the development a rather smooth process. You would still need a good understanding of C and Java Native Interface (JNI). The intent here is not to teach you C or JNI but rather the development and deployment process. It is assumed that the user is familiar with C and JNI development under eclipse.

The very first step is to write a java class similar to the one developed in the previous section. Except the fact, that we are using native method over here. Here is an example,

Listing 4.29

```
import org.numEclipse.toolbox.Toolbox;
import org.numEclipse.linalg.*;
public class NativeToolbox implements Toolbox {
  private static LinearAlgebra alg;
  public native double sqr(double x);
  public native String msg(String s);
  public static void setAlgebra(LinearAlgebra a) {
    alg = a;
  }
  public static double nativeSquare(double x) {
    System.loadLibrary("NativeToolbox");
    NativeToolbox nat = new NativeToolbox();
    return nat.sqr(x);
  }
  public static String nativeMesg(String s) {
    System.loadLibrary("NativeToolbox");
    NativeToolbox nat = new NativeToolbox();
```

```
18  return nat.msg(s);
19 }
20 }
```

Similar to the previous section, this class implements the interface *"Toolbox"*, declares an attribute *"alg"* of type *"LinearAlgebra"* and defines a setter method for this variable. Here we declare two native functions on lines 5 and 6. numEclipse cannot make a direct call to native functions so we define wrapper functions in the class with the names starting with keyword *"native"*. This naming convention is used by the interpreter to determine the nature of the function. The notion of wrapper function is developed with the view that some conversion will be required for the arguments before and after calling a native method. We previously mentioned that a double or complex argument in a function call is converted to a matrix. In this case, when you name the function starting with native, it will not perform this conversion to matrix. This design decision is made with the view that a lot of time a developer will be more interested in integrating existing C libraries rather than writing new ones. In that case, it makes more sense to keep the function arguments as it is, because existing libraries are unlikely to be using the matrix object defined in numEclipse.

As a next step, we generate the C header file *"NativeToolbox.h"* using *"javah"* utility on the last java class. Here is the generated code.

Listing 4.30

```
/* DO NOT EDIT THIS FILE - it is machine generated */
#include <jni.h>
/* Header for class NativeToolbox */
#ifndef _Included_NativeToolbox
#define _Included_NativeToolbox
#ifdef __cplusplus
extern "C" {
#endif
/*
* Class: NativeToolbox
* Method: sqr
* Signature: (D)D*/
JNIEXPORT jdouble JNICALL Java_NativeToolbox_sqr(JNIEnv *, jobject,
jdouble);
/*
* Class: NativeToolbox
* Method: msg
* Signature: (Ljava/lang/String;)Ljava/lang/String;
*/
JNIEXPORT jstring JNICALL Java_NativeToolbox_msg(JNIEnv *, jobject, jstring);
```

```
#ifdef __cplusplus
}
#endif
#endif
```

The C header file is not supposed to be modified as it says in the comments within the file. Next, we write a "*.def*" file as follows.

Listing 4.31

```
EXPORTS
Java_NativeToolbox_sqr
Java_NativeToolbox_msg
```

It basically lists the name of the functions exposed for export. Then, we write the C program *"NativeToolbox.c"* implementing the C functions in the header file.

Listing 4.32

```
//File: NativeToolbox.c
#include <jni.h>
#include "NativeToolbox.h"
JNIEXPORT jdouble JNICALL Java_NativeToolbox_sqr
   (JNIEnv *env, jobject obj, jdouble d) {
     return d * d;
}
JNIEXPORT jstring JNICALL Java_NativeToolbox_msg
   (JNIEnv *env, jobject obj, jstring s) {
     return s;
}
```

Finally, we write the following makefile to generate the *"dll"* file.

Listing 4.33

```
all : NativeToolbox.dll
NativeToolbox.dll : NativeToolbox.o NativeToolbox.def
gcc -shared -o NativeToolbox.dll NativeToolbox.o NativeToolbox.def
NativeToolbox.o : NativeToolbox.c NativeToolbox.h
gcc -I"C:\\Program Files\\Java\\jdk1.5.0_09\\include" -I"C:\\Program
Files\\Java\\jdk1.5.0_09\\include\\win32" -c NativeToolbox.c -o NativeToolbox.o
NativeToolbox.h : NativeToolbox.class
C:\Program Files\Java\jdk1.5.0_09\bin\javah -jni NativeToolbox
clean:
-del NativeToolbox.h
-del NativeToolbox.o
```

Now, the code is ready to be deployed. We will follow similar steps to those used in the previous section. We will export the java class to a *"jar"* file and add the *"jar"* and *"dll"* files both to the library preference of numEclipse. You are now ready to test the native functions, as follows.

Listing 4.34

```
>> nativeSquare(10)
ans =
100.0000
>> nativeMesg('Hello')
ans =
Hello
```

So this completes the chapter. Here we learnt how to develop and deploy a library of functions in java and C.

5

Architecture

This chapter presents the design and architecture of numEclipse. It is intended for people interested in using this tool as a research vehicle. Reading this chapter will also be beneficial for programmers writing a toolbox. Those users not interested in programming or writing engine extension could skip this chapter. Like any interpreter the design of this application can be divided into two major pieces. The front-end which deals with the scanning and parsing of the input program and back-end which actually executes the code. This chapter is also organized along the same lines. Here, we also show how to develop and deploy a custom execution engine. A number of open source tools and mathematical libraries are used in the development of numEclipse. We will also talk about their role and interfaces with the application.

5.1 Front-end

An interpreter front-end performs two tasks. It scans the input to identify the tokens and it parses the input into an Abstract Syntax Tree (AST). Traditionally, compiler/interpreter developers have used lex and yacc like tools to generate the lexer and parser programs from the language specification, i.e., grammar. We used the similar approach and rather than writing the lexer and parser from the scratch, we used SableCC[1]. This amazing tool is based on a sound object-oriented framework. Given a grammar, it generates tree-walker classes based on an extended visitor design pattern. The interpreter is built by implementing the actions on the AST nodes generated by SableCC. In the following, we present a code snippet showing how the interpreter is actually invoked by the application.

---

[1] http://www.sablecc.org

Listing 5.1

```
Reader strReader = new StringReader(input);
Lexer lexer = new Lexer(new PushbackReader
            (new BufferedReader(strReader)));
Parser parser = new Parser(lexer);
Node ast = parser.parse();
ast.apply(interpreter);
```

Writing the language specification (Grammar) is the most complicated part of a language development. Once the grammar is solidified, generation of a Lexer and Parser classes using SableCC is just a click of a button. Most of the effort in developing this application involved writing the class *"Interpreter"*. It extends *"DepthFirstAdapter"* class which is generated by SableCC. This adapter class is the AST tree walker mentioned earlier.

5.2 Back-end

So what constitutes the back-end? The back-end is where the action happens. The back-end starts from the class which extends *"DepthFirstAdapter"* (i.e., Interpreter) class. This tree-walker class has the action code for each node encountered in the AST during parsing of the input program. Here is the list of actions that happen in this class.

1. Creating variables,
2. Storing variables in a symbol table,
3. Evaluating expressions,
4. Executing statements, and
5. Calling functions.

The m-script, as mention in earlier chapters, does not require you to declare a variable. You can just start working with a variable and the interpreter will figure out the value and type of the variable. This poses some implementation challenges but provides a lot of flexibility to the end-user. In the previous chapter on programming, we referred to interface class *"LinearAlgebra"*. It provides methods to create different type of variables. The implementation of the interface classes constitutes the execution engine. These classes not only provide the methods to create the different variables but also provide the basic arithmetic operations to evaluate complex expressions. Good understanding of the functions of these classes is essential in order to implement an alternative execution engine. The symbol table is implemented as a hash table. In fact, it contains three hash tables for ordinary symbols, global symbols and constants. *"Symbol"* is another object which is used to wrap any variable before it is stored into the symbol table. Each instance of the interpreter window gets its own symbol table, so you can only see the symbols in the memory view which are tied to the active interpreter. The symbol table extends the

"Observable" class so that the memory view could register as an *"Observer"* and show the changes as they happen. The Symbol class implements the Serializable interface, so that the variables could be easily saved and retrieved from a file. This enabled us to save a session in a file.

Expression evaluation depends entirely on the basic arithmetic operations on different data types supported by numEclipse. As mentioned earlier, these operations are defined within the implementation classes which form the execution engine. Statements are discussed previously in chapter 3. They are very similar to any other programming languages like C or FORTRAN. The correct execution of these statements is the responsibility of the Interpreter class. This functionality is fixed and cannot be modified for obvious reasons. numEclipse has a number of built-in functions and it offers the ability to integrate user-defined functions. On the start-up, the application loads all the m-script and java functions into a library. All functions, built-in or user-defined, are loaded through a common mechanism using Java Reflection APIs. The library manager also keeps track of the *"dll"* files added by the user as described in the previous chapter. Java reflection is known to be slow in loading a class/method but in numEclipse all functions are pre-loaded in a hash table so the cost of calling the functions is not so high. The library manager maintains a precedence rule for the function calls. It looks up a function in the following order.

1. user-defined m-script function in the numEclipse project,
2. user-defined java function within the referenced projects in the eclipse workspace,
3. user-defined m-script function added to the preferences,
4. user-defined java function added to the preferences,
5. built-in java function and
6. built-in m-script function.

At the moment, this order is fixed but in future we might allow the user to change this precedence rule through preferences. This completes an overview of the interpreter back-end for more insight one needs to go over the source code.

5.3 User Interface

The very first user interface of numEclipse never saw daylight. It was built on Java Swing. It was quickly realized that it does not really serves the objectives of this project. The intention behind numEclipse is not just an interpreter but rather a comprehensive development environment for scientific computing. However, MATLAB or GNU Octave do provide the possibility to add functions in other programming languages but they do not provide any integration with the development tools as such. We decided to re-write numEclipse as an eclipse plug-in and this approach opened up a whole new world of opportunities for us. In previous chapters, we showed how to write a java or C function within eclipse and how to quickly test and deploy them with numEclipse. This seamless integration would not have been possible without the eclipse platform.

We decided to follow the software engineering approach for scientific application development. So, we introduced the notion of a numEclipse project. This gives a project oriented and role based development of scientific application. We created a new perspective for the project development. We also added a wizard to create a new project. The perspective contains three new components, i.e., interpreter window (editor), memory view and history view. The interpreter window is basically an editor in eclipse's terms. We do not know of any other interpreter implementation within eclipse so we developed the interpreter (editor) from scratch. The design of this interpreter is still in development and there are a lot of opportunities for improvement. The memory and history views were rather easy to develop. They use the observer-observable design pattern to update their corresponding information. At the moment the interpreter window is very much hocked to the actual interpreter and we are trying to come up with a better design to introduce separation of concerns. This might set precedence for future interpreter plug-ins for eclipse. Another user interface component is the numEclipse preferences as we saw in the previous chapters. This enables us to define numEclipse related configurations like constants, libraries and gnuplot.

5.4 Gnuplot Interface

Our initial intent was to write Java2D/Draw2D based plotting APIs. But we quickly realized that this would be an enormous task and that there is no point in reinventing the wheel. There are already a number of open source projects providing excellent APIs for plotting. Our objective was to choose something similar to MATLAB. We started looking at PLPlot first, it is a set of plotting functions written in C. This project also provides java binding to the C functions Unfortunately, this project is more geared towards linux/UNIX users. We initially compiled the JNI enabled dll on WindowsXP and came across a lot of problems. PLPlot functions have their own windows management; once a graph is plotted by a java program through binding, it has no control over the plot window. Also we discovered that you could only have one plot at a time which is not acceptable for our purpose. Finally, we decided to take the approach of Octave and provided an interface to gnuplot. It is an excellent tool for scientific plotting. It has been developed over a long period of time. We are using version 4.0 and it is very mature and stable. We provide this interface to gnuplot as a built-in toolbox. We are hoping that some users will try to write their own toolboxes for other visualization APIs or applications.

Gnuplot is an application rather than a set of APIs. It provides a command line user interface. This posed another challenge for integration. But fortunately, gnuplot also accepts the user commands through a named pipe. Now, you would understand why we need to define the path to the gnuplot execution file within the numEclipse preference. In the following, we show the code snippet used to invoke gnuplot and create a link.

Listing 5.2

```
Process p = Runtime.getRuntime().exec(gnuplot);
PrintStream out = new PrintStream(p.getOutputStream());
```

Once a link is established, to send a command to gnuplot we use the following.

Listing 5.3

```
String command = ....
out.println(command);
out.flush();
```

So you see the integration with gnuplot is very straightforward. Most of effort involved writing methods which translated numEclipse commands into gnuplot commands. On top of that, we had to store the data and states in between the commands. To store the temporary plotting data, we create temporary files in user area allocated by the operating system. These files are short lived and scratched at the end of the session. In order to get more information, one needs to walk through the source code, i.e., org.numEclipse.toolbox.Plot.

5.5 Execution Engine

In this section, we will show how to develop and deploy an execution engine. The intent is to show the process with a simple example rather than building a sophisticated engine. Let's give a different meaning to matrix computation. We re-define the matrix addition, subtraction and multiplication using the following formulae.

$$A \oplus B = A + B \bmod N$$
$$A \Theta B = A - B \bmod N$$
$$A \otimes B = A \times B \bmod N$$

The *"mod"* stands for the modulo operation. The result of *"a mod b"* is the remainder term, when *"a"* is divided by *"b"*. The symbols ⊕, Θ, ⊗ are used here only to distinguish, otherwise the arithmetic operator symbols remain the same within numEclipse. A good implementation of these operators will take a lot of effort. We will make a quick implementation to prove the concept.

We use the *"DefaultMatrixImpl"* Class which implements the interface *"IMatrix"* as described on the project website (http://www.numeclipse.org/interface.html). We refactor the class and copy it as *"ModuloMatrixImpl"* Class. Then we modify the following methods.

Listing 5.4

```
Public IMatrix mult(IMatrix m) {...}
Public IMatrix plus(IMatrix m) {...}
Public IMatrix minus(IMatrix m) {...}
```

We do not show the code of these methods as the change is extremely simple. We apply the following utility function on each element of the resultant matrix before we return the value.

Listing 5.5

```
private IComplex modulo10(IComplex z) {
  double re = z.getReal();
  double im = z.getImag();
  IComplex z = new DefaultComplexImpl(re % 10, im % 10);
  return z;
}
```

Then, we refactor the *"DefaultLinearAlgebraFactory"* class and copy it as *"ModuloLinearAlgebraFactory"* class. Then, we modify the following methods as shown.

Listing 5.6

```
public IMatrix createMatrix(IComplex[][] c) {
  return new ModuloMatrixImpl(c);
}

public IMatrix createMatrix(double[] d1, double[] d2) {
   return new ModuloMatrixImpl(d1, d2);
}

public IMatrix createMatrix
         (double[][] d1, double[][] d2) {
   return new ModuloMatrixImpl(d1, d2);
}

public IMatrix createMatrix(IMatrix m) {
   return new ModuloMatrixImpl(m);
}

public IMatrix createMatrix(int m, int n) {
   return new ModuloMatrixImpl(m, n);
}
```

```
public IMatrix createMatrix(String[][] str) {
   return new ModuloMatrixImpl(str);
}

public IMatrix createMatrix
          (Hashtable hash, int m, int n) {
   return new ModuloMatrixImpl(hash, m, n);
}

public IMatrix createMatrix(BigDecimal[][] b) {
   return new ModuloMatrixImpl(b);
}
```

The change is very simple, all we did is change the call to the new constructor in the class *"ModuloMatrixImpl"*. Notice that the change is minimal; we did not modify any other data type. We also did not modify the structure of the matrix data type. We only modified the way addition, multiplication and subtraction of two matrices work.

In order to deploy this simple engine, we export these two classes into a jar file (say *modulo.jar*). Then, add the jar file to the library with numEclipse Preferences. The application will ask you to restart the workspace, allow the application to automatically restart. Right-click anywhere in the interpreter window, a pop-up menu will appear, select the new execution engine. Now you are ready to test the changes. In the following, we show some calculations with default engine and then we show we show the results with new engine.

Listing 5.6 (Default implementation)

```
>> A = [4 5; 3 9];
>> B = [3 0; 7 9];
>> A + B
ans =
7.00005.0000
10.0000   18.0000
>> A - B
ans =
1.00005.0000
-4.00000  0.0000
>> A * B
ans =
47.0000   45.0000
72.0000        81.0000
```

Listing 5.7 (Modulo implementation)

```
>> A = [4 5; 3 9];
>> B = [3 0; 7 9];
>> A + B
ans =
7.00005.0000
0.00008.0000
>> A - B
ans =
1.00005.0000
-4.00000  0.0000
>> A * B
ans =
7.00005.0000
2.00001.0000
```

Once the new engine is loaded through the preferences, you can switch back and forth from one engine to another just with a click of a mouse button. However, there is a catch, a variable created with the default engine will use the arithmetic operations defined in the default engine. So in other words, just because you switched the engine does not mean that you will be able to apply the new operations with existing variables in the workspace memory. You should clear the memory and create the variables again to use the new operations. In future, we might add a utility to convert the variables as you change the engine. In this section, we showed how to create a simple engine and how to deploy it.

6

Plotting

Data visualization or plotting is an important part of scientific computing. In this chapter, we will demonstrate how to invoke the plotting capabilities of gnuplot through numEclipse. It is assumed that the user has installed the latest version of gnuplot and numEclipse preferences are configured to point to the gnuplot as shown in the installation guide on the numEclipse project website (http://www.numeclipse.org/install.html).

6.1 Simple Function Plot (fplot)

The "fplot" method is used to plot a mathematical function. It is very different from the MATLAB or Octave implementation. It can only plot the functions recognized by gnuplot. Fortunately, gnuplot has a good collection of built-in functions.

```
>> fplot('sin(x)')
```

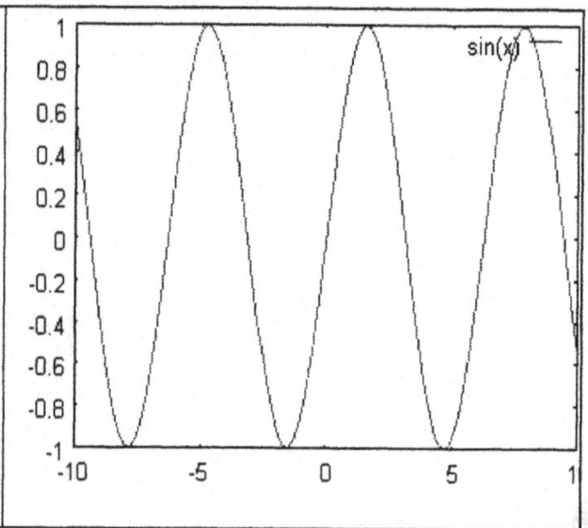

Figure 6.1

Unlike MATLAB, the function name is not enough e.g. *"sin"*. We also need to provide the arguments to the function i.e. "sin(x)". However, the numbers of supported functions are limited but it does not stop a user to define new functions using algebraic combination of existing functions. For example,

```
>>fplot(
'(sin(x)-cos(x))/
(sin(x)+cos(x)',
'[-pi:pi]',
'[-150:150]')
```

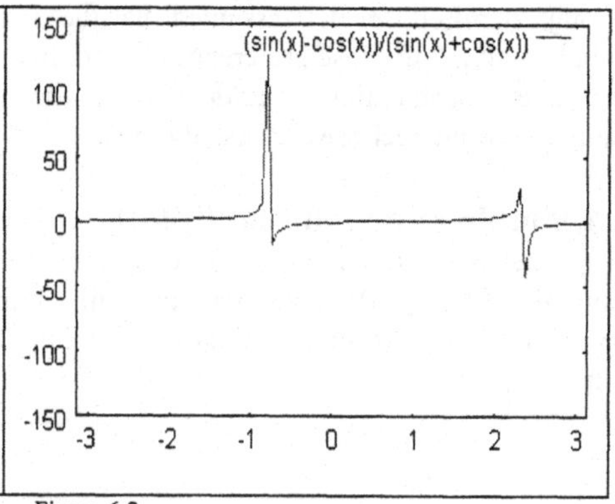

Figure 6.2

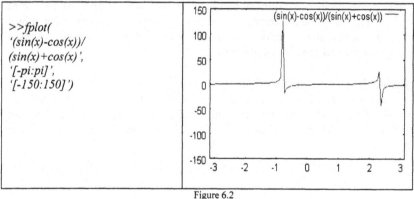

In this example, we also define the x-range and y-range of the function plot. Please note that the range arguments are strings and not arrays unlike MATLAB. The syntax for using fplot is given in the following.

fplot ('function-name(argument)')

fplot ('function-name(argument)', '[xmin:xmax]')

fplot ('function-name(argument)', '[xmin:xmax]', '[ymin:ymax]')

fplot ('function-name(argument)', '[xmin:xmax]', '[ymin:ymax]', 'format')

The last argument of the fplot method controls the plot format. The first character represents the color and second character represents the line type of the plot. The same format convention is used in the rest of the plotting functions discussed in this chapter.

Table 6.1

Color	**Line type**
r (red)	*
g (green)	+
b (blue)	-
m (magenta)	.
	o
	x

In the following, we repeat the last example with a format option.

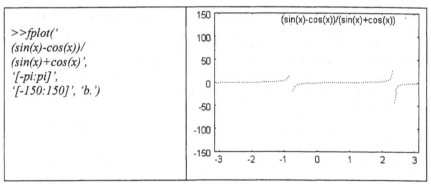

>>fplot('
(sin(x)-cos(x))/
(sin(x)+cos(x)',
'[-pi:pi]',
'[-150:150]', 'b.')

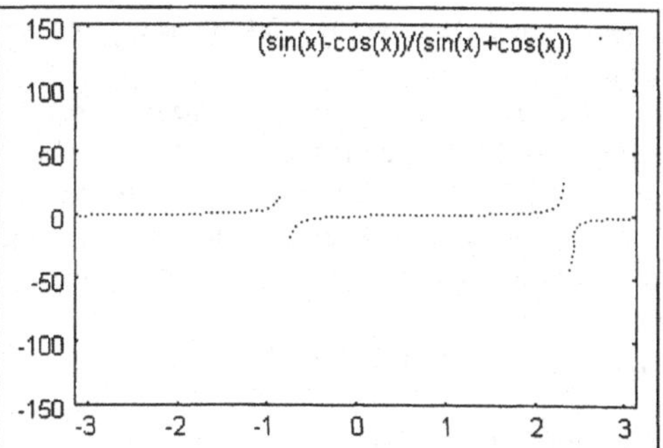

Figure 6.3

The following example shows how to annotate a plot with a title and labels. We also show how to overlay multiple plots and how to add a grid to a plot.

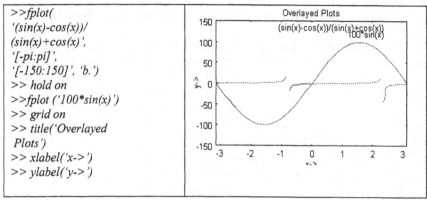

```
>>fplot(
'(sin(x)-cos(x))/
(sin(x)+cos(x)',
'[-pi:pi]',
'[-150:150]', 'b.')
>> hold on
>>fplot ('100*sin(x)')
>> grid on
>> title('Overlayed
Plots')
>> xlabel('x->')
>> ylabel('y->')
```

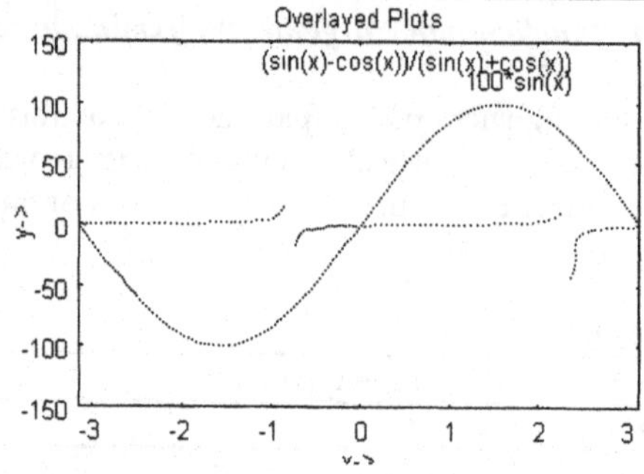

Figure 6.4

Command *"hold on"* holds the current plot and axes while the new plot is rendered. The new plot cannot modify the current properties of the plot for example the axes range cannot be changed. Command *"hold off"*, cancels the hold and the new plot will erase the existing plot including the axes properties. *"grid on"* shows the grid lines in the current plot and all the next plots until it is disabled using the *"grid off"* command.

Commands *xlabel('label')*, *ylabel('label')* and *zlabel('label')* place the *'label'* string besides the corresponding axis (x-, y- or z-axis). Command *title('label')*, places the label string on the top of the plot. The last example shows the usage of these annotation functions. These common functions are also applicable to all the plotting functions discussed in the following.

6.2 Two-Dimensional Plots

6.2.1 Basic plot (plot)

The most commonly used plot method is *"plot"*. *plot(X)* will plot the elements of *X* versus their index if *X* is a real vector. If *X* is complex vector then it will plot the imaginary component of vector *X* versus the real components. In case *X* is a real matrix then *plot(X)* will draw a line plot for each column of the matrix versus the row index. In case of complex matrix it will plot the imaginary components of the column versus their corresponding real components. *plot(X, Y)* will plot elements of vector *Y* versus elements of vector *X*. In case of complex vectors the imaginary components will be ignored and only real part of the vector elements will be used for plotting. We can add one more string argument to define the format of the plot, e.g. *plot(X, "g+")*, *plot(X, Y, "bx")*. The format is a two letter string where the first letter represents the color and second defines the line style. The possible values for color and line style are described in the previous section. A simple plot example is given in the following.

```
>> hold on;
>> t = 0:0.01:1;
>> x = sin(6*pi*t);
>> plot(t, x, "b+");
>> y = cos(6*pi*t);
>> plot(t, y, "rx");
```

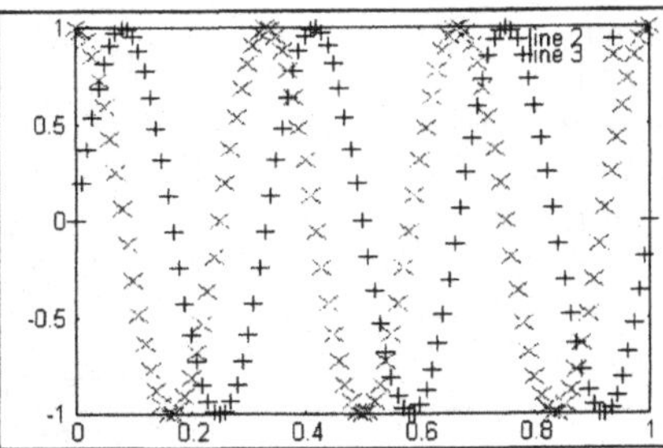

Figure 6.5

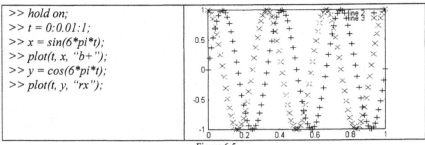

The above plot shows the points corresponding to sin function as blue plus "+" symbols and points for *cos* function as red cross "*x*" symbols. The range of each axis could be redefined using the following common commands.

axis([xmin, xmax])

axis([xmin, xmax, ymin, ymax])

axis([xmin, xmax, ymin, ymax, zmin, zmax])

"xmin" :	***minimum value of x axis***
"xmax" :	***maximum value of x axis***
"ymin" :	***minimum value of y axis***
"ymax" :	***maximum value of y axis***
"zmin" :	***minimum value of z axis***
"zmax" :	***maximum value of z axis***

In the following example, we show the last example with re-defined axes.

```
>> hold on
>> t = 0:0.01:1;
>> x = sin(6*pi*t);
>> plot(t, x, “b+”);
>> y = cos(6*pi*t);
>> plot(t, y, ``rx");
>> axis([0.4, 0.6,
-0.5, 0.5]);
```

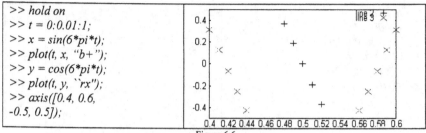

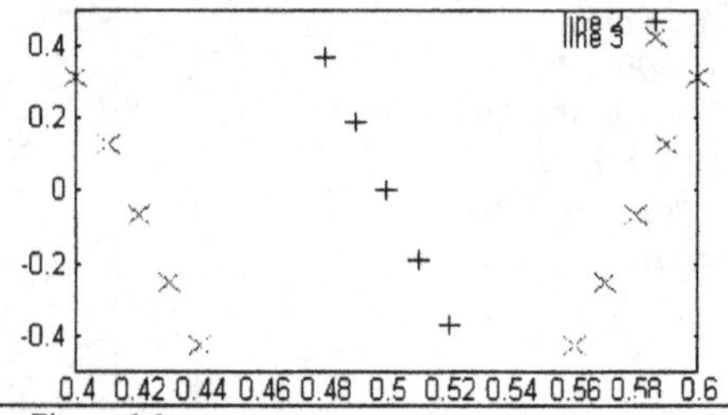

Figure 6.6

In the absence of a format option, the plot function draws a plot using an interpolated line as shown in the following example.

```
>> x = [-1:0.01:1];
>> y = x.^2;
>> plot(x, y);
```

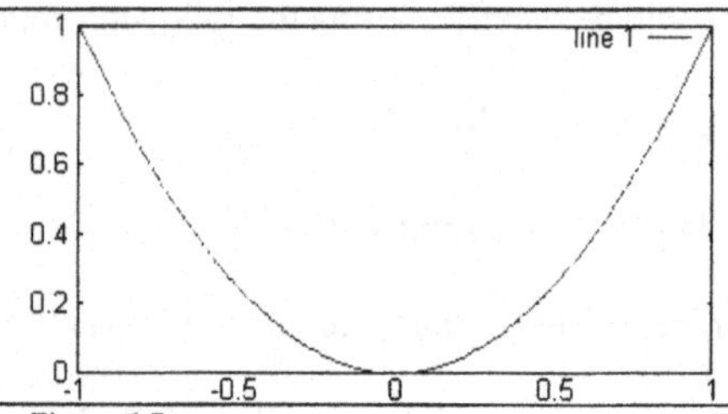

Figure 6.7

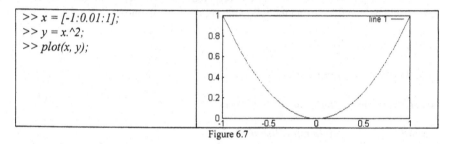

6.2.2 Logarithmic plot (plot)

The logarithmic plot functions (*semilogx, semilogy, loglog)* essentially work in the same manner as the *plot* function. They use the same argument format as the *plot* function. They differ in the way their axes are scaled.

semilogx uses log (base 10) scale for the X-axis.
semilogy uses log (base 10) scale for the Y-axis.
loglog uses log (base 10) for both X- and Y-axis.

These plots are used to view the trends in the dataset where the values are positive and there are large variations between data points. For example, an exponentially distributed dataset will show a linear relationship on a logarithmic plot.

```
>> x = [-1:0.01:1];
>> y = exp(x);
>> plot(x, y);
>> grid on
```

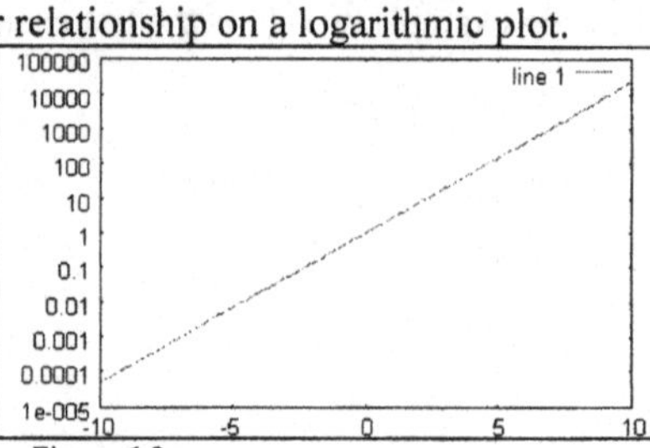

Figure 6.8

Note the equi-distant horizontal grid lines even though the difference in the *y* values is exponent of *10*.

6.2.3 Polar plot (polar)

The polar plot method *"polar (theta, rho)"* draws a plot using polar coordinates of angle *theta* (in radians) versus the radius *rho*. Like the previous methods, we can also add a third String argument defining the format of line color and style as described previously. Note, how the horizontal and vertical grid lines change into radial and elliptical grid lines.

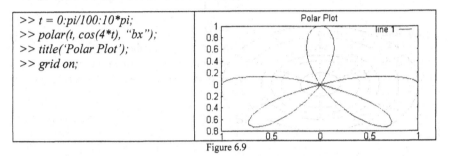

```
>> t = 0:pi/100:10*pi;
>> polar(t, cos(4*t), "bx");
>> title('Polar Plot');
>> grid on;
```

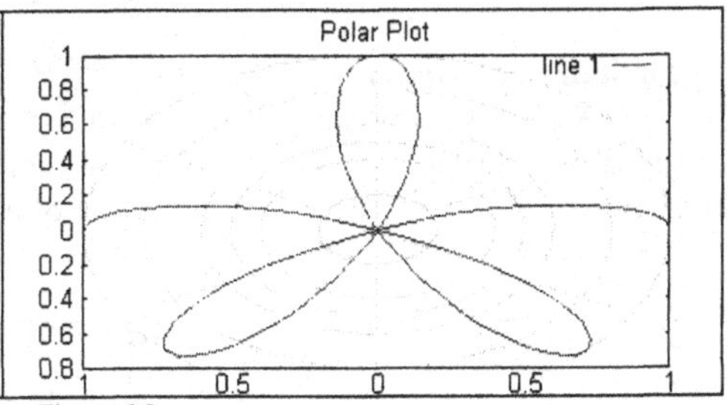

Figure 6.9

6.2.4 Bar chart (bar)

The bar chart method bar(X) makes a bar graph of elements of vector X as shown in the following example.

```
>> bar(1:10)
```

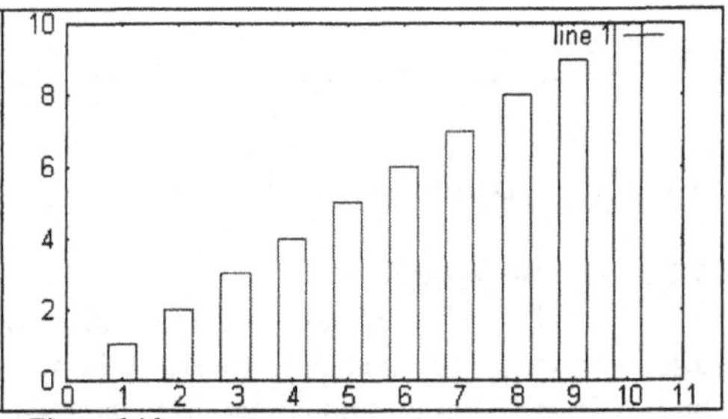

Figure 6.10

Method *bar(X, Y)* draws a bar graph of elements of vector *Y* versus the elements of vector *X*, provided elements of *X* are increasing in the ascending manner.

6.2.5 Histogram plot (hist)

The histogram plot is commonly used in any scientific presentation. The method *"hist(X)"* draws a histogram of the vector data *X*. It divides the data into *10* bins between the minimum and maximum values of *X*.

```
>> x = rand(1, 10)
x =
0.6356 0.2867 0.4402 0.3390
0.2066 ...
0.1622 0.1239 0.6027 0.7048
0.8545
>> hist(x)
```

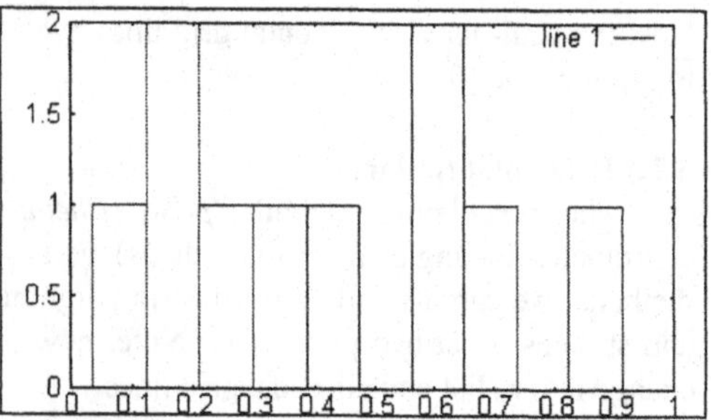

Figure 6.11

6.2.6 Stairs plot (stair)

The stairs plot is not a very common plot. The method *stairs(X)* makes the stairstep plot of the elements of vector *X* versus the element index. Similarly the method *stairs(X, Y)* draws the stairstep plot of elements of vector *Y* versus elements of vector *X*.

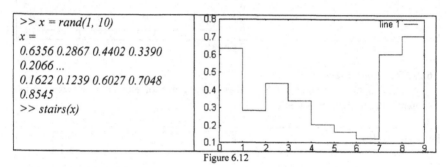

```
>> x = rand(1, 10)
x =
0.6356 0.2867 0.4402 0.3390
0.2066 ...
0.1622 0.1239 0.6027 0.7048
0.8545
>> stairs(x)
```

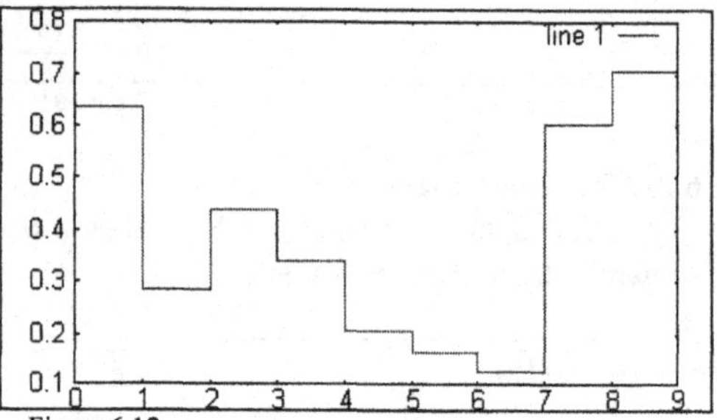

Figure 6.12

6.3 Three-Dimensional Plots

6.3.1 Line plot (plot3)

The 3D version of the previously discussed plot method is *"plot3"*. It takes three vectors as arguments representing the *x*, *y* and *z* components of 3-dimensional line. A fourth String argument could be added to define the format of the line in the same manner as plot method. At this point, this method does not support matrix arguments, unlike MATLAB.

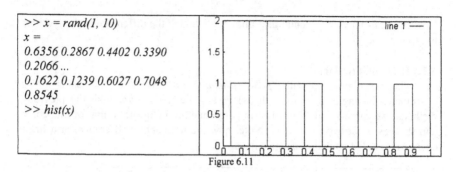

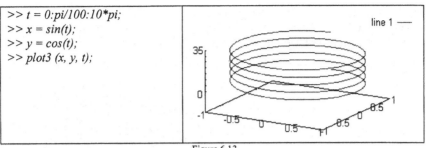

```
>> t = 0:pi/100:10*pi;
>> x = sin(t);
>> y = cos(t);
>> plot3 (x, y, t);
```

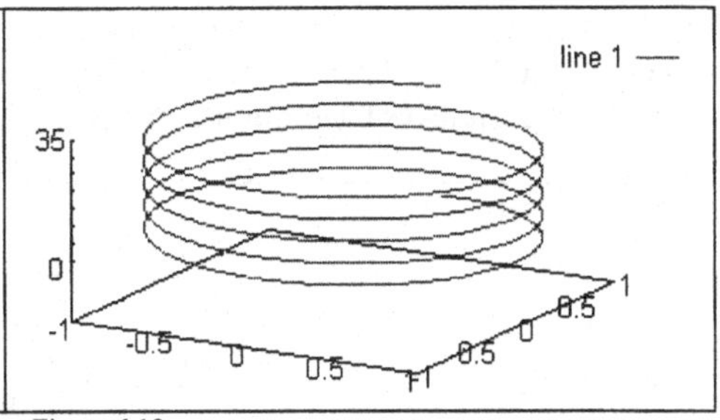

Figure 6.13

6.3.2 Mesh surface plot (mesh, meshc, contour)

The mesh surface plot is a popular 3D plot for scientific visualization. The method *"mesh(X, Y, Z)"* plots a meshed surface defined by the matrix argument *Z* such that vectors *X* and *Y* define the *x* and *y* coordinates. This method without the first two arguments will use the row and column index of the matrix *Z* to plot the mesh.

```
>> x = -2:0.05:2;
>> y = x;
>> for i=1:length(x)
  for j=1:length(y)
    z(i, j) = 2 *   (x(i)^2 + y(j)^2) *
              exp(-x(i)^2 - y(j)^2);
  end;
end;
>>mesh(x, y, z)
```

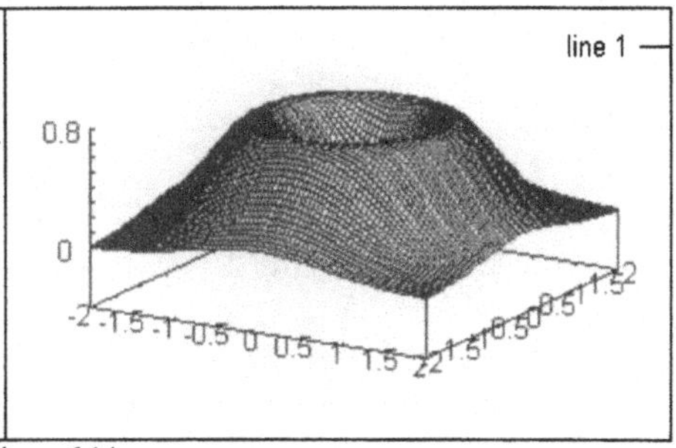

Figure 6.14

A variation of this method is *meshc(X, Y, Z)*. This method enhances the mesh plot by adding contours. In the absence of first two arguments it works the same way as method *mesh*.

```
>> meshc(x, y, z);
```

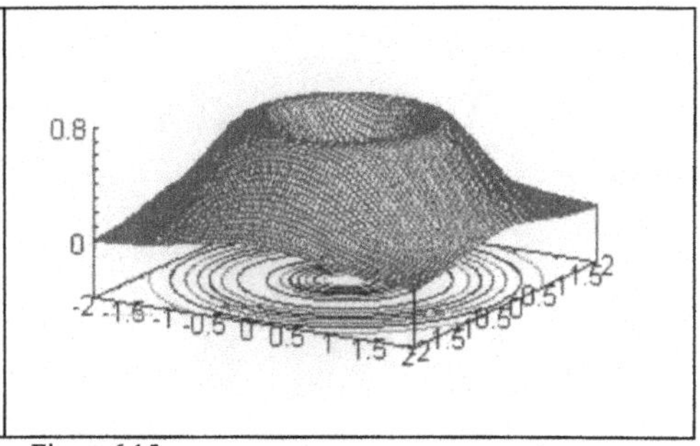

Figure 6.15

Sometimes we are only interested in the topography of the surface rather than the complete mesh plot. There is another method contour, which plots only the contour of the surface. It takes the same arguments as mesh or meshc.

>> *contour(x, y, z);*	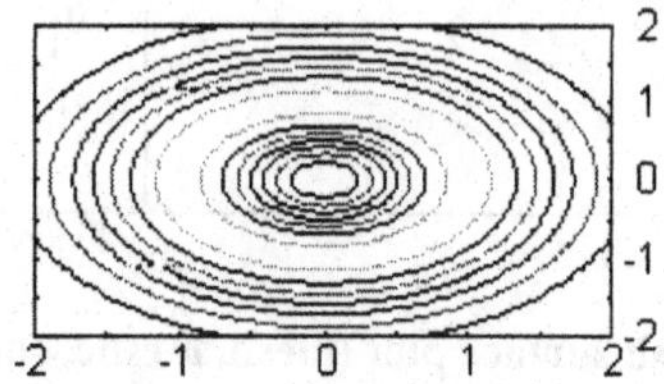

Figure 6.16

Part II

In this part, we consider some classical methods of numerical analysis. The part includes the following topics: finding roots of transcendental equations, solving systems of linear equations, finding eigenvalues, finite difference schemes for ODE, and data representation and approximation of functions. The basic ideas of numerical techniques and their practical implementation are presented. The numerical algorithms and their implementations using numEclipse are also presented. Numerous exercises are provided that can be used in practical assignments.

7

Solving Nonlinear Equations

Finding a solution for an equation of the form $f(x)=0$ is a frequently-encountered problem. It is not as easy to find roots of this equation as it might at first appear. Exact solutions are only possible if $f(x)$ is a polynomial of degree $n \leq 9$. By "exact solution," we mean there is some procedure to calculate a root by using the parameters of the equation (as in the equation $ax^2+bx+c=0$). In what follows, the term "root of an equation $f(x)=0$" will be used interchangeably with "zero of $f(x)$".

The root finding procedure is based on an iterative process or, in other words, on a process of successive approximations. An iterative process has four stages:

1. Root localization – that is, finding an interval which contains the only root.
2. Choosing an initial value, x_0, or, in other words, letting x_0 be an approximate value of x^*.
3. Employing some procedure k times to obtain a set of approximations x_k, starting from x_0.
4. Finding out how close every approximation, x_k, is to the exact solution (in terms of prescribed accuracy ε_p). If some approximation is in the ε_p-neighborhood of x^*, then the iterative process is complete.

The last point will be satisfied only if:

$$\lim_{k \to \infty} x_k = x^*,$$

that is, approximations x_k converge to x^*. That is why a great deal of attention will be given to conditions wherein an iterative process converges.

7.1 Calculation of Roots with the use of Iterative Functions

In what follows, we will consider another approach to calculate roots. An equation $f(x)=0$ may be represented in the following equivalent form:

$$x = g(x), \tag{7.1}$$

where $g(x)$ is an iterative function. Thus, $f(x^*)=0$ is consistent with $x^*=g(x^*)$. It turns out that form (7.1) is more appropriate for the construction of iterative algorithms. An iterative process is constructed in the following manner: let us set the value of x_0 and the succeeding approximations will be calculated by the formula

$$x_{k+1} = g(x_k), \ k=0, 1, \ldots. \tag{7.2}$$

Figure 7.1 shows the geometrical interpretation of the convergent iterative process (7.2).

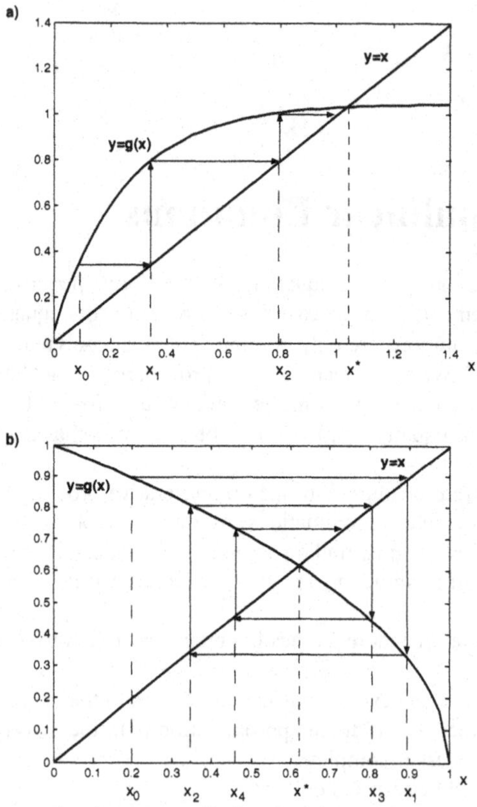

Figure 7.1 Successive approximations generated by iterative scheme (7.2): a) g'(x)>0 and b) g'(x)<0.

The following theorem explains conditions which provide convergence of the iterative process (7.2).

Theorem 7.1

Let $I=[a, b]$ be a closed, bounded interval, and let $g(x)$ be a function from I to I. In addition, $g(x)$ satisfies the Lipschitz condition

$$|g(y) - g(z)| \le L|y - z|, 0 \le L \le 1,$$

for arbitrary points y and z in I. Let $x_0 \in I$ and $x_{k+1}=g(x_k)$. Then the sequence $\{x_k\}$ converges to the unique solution $x^* \in I$ of the equation $x=g(x)$.

One can derive an estimate of the error of the kth approximant, which depends only on the two successive approximants and the Lipschitz constant. Because $|x_{k+1}-x_k|=|g(x_k)-g(x_{k-1})|$, we can conclude from the Lipschitz condition that for all k, $|x_{k+1}-x_k| \le L|x_k-x_{k-1}|$. Then

$$|x_k - x^*| \le \frac{L}{1-L}|x_k - x_{k-1}|. \tag{7.3}$$

This gives us a bound on the error of the kth approximant. It should be noted that if L is close to one, convergence is very slow. So, a good part of our effort will be concentrated on how to construct an iterative process which converges rapidly.

There are a lot of methods for solving an equation $f(x)=0$. We will consider only basic approaches. The iterative processes to be considered may be categorized as follows:

- One-point iteration:

$$x_{k+1} = g(x_k), k=0, 1, \ldots, \tag{7.4}$$

- Multi-point iteration:

$$x_{k+1} = g\left(x_k, \beta_1(x_k), \ldots, \beta_n(x_k)\right) k=0, 1, \ldots, \tag{7.5}$$

where $\beta_1(x), \ldots, \beta_n(x)$ are some functions.

- One-point iteration with memory:

$$x_{k+1} = g\left(x_k, x_{k-1}, \ldots, x_{k-n}\right), k=n, n-1, \ldots. \tag{7.6}$$

Now, the term "rapid convergence" should be explained in more detail. To do this, we take the iterative process (7.4) as an example. To begin with, let us introduce the error of kth approximation as $e_k=x_k-x^*$, then $x_k=x^*+e_k$ and $x_{k+1}=x^*+e_{k+1}$. After substituting these expressions in the iterative process (7.4), we can expand $g(x^*+e_k)$ in terms of e_k in the neighborhood of x^* (assuming that all derivatives of $g(x)$ up to order p are continuous):

$$x^* + e_{k+1} = g(x^* + e_k) = g(x^*) + g'(x^*)e_k + \frac{1}{2}g''(x^*)e_k^2 + \ldots + \frac{1}{p!}g^{(p)}(y_k)e_k^p,$$

$$y_k \in [x_k, x^*].$$

Taking into account that $x^*=g(x^*)$ we arrive at

$$e_{k+1} = g'(x^*)e_k + \frac{1}{2}g''(x^*)e_k^2 + \ldots + \frac{1}{p!}g^{(p)}(y_k)e_k^p. \tag{7.7}$$

If $g^{(n)}(x^*)=0$ for $n=1, \ldots, p-1$ and $g^{(p)}(x^*)\neq 0$, then an iterative function, g, has an order, p, and thus it follows immediately from equation (7.7) that

$$e_{k+1} = Ce_k^p, C=\text{const}\neq 0, \tag{7.8}$$

where p is called the order of the sequence $\{x_k\}$ and C is called the asymptotic error constant. When $p=1$, an iterative sequence has linear convergence to x^*, while with $p>1$, convergence is superlinear. It is clear that if e_k is sufficiently small, then with $p>1$ an iterative sequence may converge very rapidly.

Next, we need to discuss how to estimate how closely the kth approximation approaches the exact solution x^*. To do this, we can use inequality (7.3). If $x_k \in [x^* - \varepsilon_p, x^* + \varepsilon_p]$, then the following conditions are fulfilled for absolute and relative error estimation, respectively:

$$|x_k - x_{k-1}| \le \varepsilon_p \text{ or } \frac{|x_k - x_{k-1}|}{|x_k|} \le \varepsilon_p . \tag{7.9}$$

In addition to conditions (7.9), usually another condition must be satisfied

$$|f(x_k)| \le \varepsilon_p .$$

This condition follows from the formulation of the problem. Therefore, an iterative process is finished when one of those conditions is fulfilled.

7.1.1 One-point iterative processes

To begin with, we will consider the case when $g'(x^*)\neq 0$. Then, as it follows from equation (7.7) that $e_{k+1}=g'(x^*)\, e_k$ and with the proviso that

$$|\varphi'(x^*)| < 1, \tag{7.10}$$

convergence is linear. In practice, one cannot use condition (7.10), but we can require that the following condition be satisfied:

$$\max_{x\in I} |\varphi'(x)| < 1,\; x^* \in I. \tag{7.11}$$

In this case, theoretical condition (7.10) is also satisfied. The simplest way to represent a source equation in form (7.1) is as follows:

$$x = x + \alpha f(x) = g(x),\; \alpha = \text{const} \neq 0, \tag{7.12}$$

and the associated iterative scheme is

$$x_{k+1} = x_k + \alpha f(x_k) = g(x_k),\; k=0,\, 1,\, \dots . \tag{7.13}$$

This is the method of fixed-point iteration. In order that condition (7.11) is satisfied at every point in I, we should determine parameter α as follows:

$$0 < \alpha < \frac{2}{\max_{x\in I} |f'(x)|} \text{ if } f'(x)<0,$$

To be more specific, we can choose

$$\alpha = \begin{cases} \dfrac{1}{\max_{x\in I} |f'(x)|} & \text{if } f'(x) < 0 \\ -\dfrac{1}{\max_{x\in I} f'(x)} & \text{if otherwise} \end{cases} .$$

The initial value x_0 should be chosen from interval I.

Convergence of the method of fixed-point iteration is not very rapid, but there is a procedure to decrease the number of iterations using the same information. Expression (7.7) for method (7.13) can be written as

$$e_{k+1} = Re_k + O(e_k^2)\,,\ R = g'(x^*)$$

or

$$x_{k+1} - x^* = R(x_k - x^*) + O(e_k^2)\,.$$

Let us consider two consecutive iterations:

$$x_k - x^* = R(x_{k-1} - x^*) + O(e_{k-1}^2)\,,$$

$$x_{k+1} - x^* = R(x_k - x^*) + O(e_k^2)\,.$$

After eliminating R, we can express x^* from those two expressions

$$x^* = \frac{x_k^2 - x_{k+1}x_{k-1}}{2x_k - x_{k+1} - x_{k-1}} + O(e_k^2) = z_{k+1} + O(e_k^2)\,. \qquad (7.14)$$

If we take z_{k+1} as an approximation to x^*, then the accuracy of this approximation is $O((e_k)^2)$ – that is, the sequence $\{z_k\}$ has convergence of the second order, as opposed to the sequence $\{x_k\}$ which has linear convergence. The transformation $\{x_k\}\rightarrow\{z_k\}$ is called Aitken's δ^2-process.

There are a lot of procedures to achieve superlinear convergence. Let the derivatives of $f(x)$ up to order s be continuous in I. Then, in the neighborhood of $x_k \in I$, $f(x)$ can be represented as a polynomial using Taylor's formula (Figure 7.2)

$$f(x) = P_s(x) + \frac{f^{(s)}(y_k(x))}{s!}(x - x_k)^s\,, \qquad (7.15)$$

$$P_s(x) = \sum_{n=0}^{s-1} \frac{f^{(n)}(x_k)}{n!}(x - x_k)^n\,,$$

where y_k lies in the interval determined by x_k and x. The next approximation to x^* can be found from the equation

$$P_s(x_{k+1}) = 0\,. \qquad (7.16)$$

By this means, expressing x_{k+1} from equation (7.16), we obtain the transformation $x_k \rightarrow x_{k+1}$, and this defines some iterative scheme. It is easy to show that the iterative sequence generated by (7.15) and (7.16) has the order $p=s$.

Next, we need to discuss how to choose an initial value x_0 which initiates a convergent iterative process. Let $x^* \in I$, with derivative $f^{(s)}(x)$ continuous in I and $f'(x)f^{(s)}(x) \neq 0$ for all $x \in I$. When conditions

$$f(x_0)f^{(s)}(x_0) > 0\,, \text{ if } s \text{ is even,} \qquad (7.17)$$

$$f'(x_0)f^{(s)}(x_0) < 0\,, \text{ if } s \text{ is odd,} \qquad (7.18)$$

are satisfied, and $\min(x^*, x_k) < x_{k+1} < \max(x^*, x_k)$, then the iterative sequence $\{x_k\}$ generated by (7.15) and (7.16) converges monotonically to x^*.

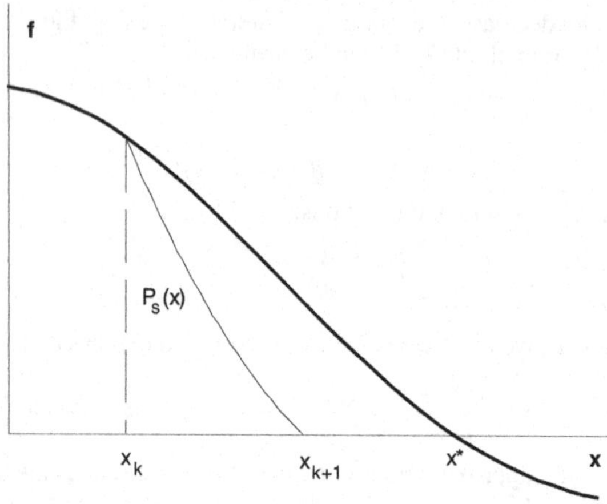

Figure 7.2 Polynomial approximation of the function $f(x)$ near the point x_k.

Now, let us consider the particular methods which follow from general approach (7.15) and (7.16). When s=2, equation (7.16) has the following form:

$$f(x_k) + f'(x_k)(x_{k+1} - x_k) = 0\,.$$

Solving this equation with respect to x_{k+1} , we obtain the following iterative scheme:

$$x_{k+1} = x_k - \frac{f(x_k)}{f'(x_k)}, \; k=0,\ 1,\ \ldots, \tag{7.19}$$

with $g(x) = x - f(x)/f'(x)$. This is the well known Newton's method and it has a simple geometrical interpretation. Function $y(x)=f(x_k)+f'(x_k)(x-x_k)$ is the tangent to the curve of $f(x)$ at the point x_k. Then approximation x_{k+1} is the intersection point of the tangent with the x-axis. To choose an initial value x_0, it is necessary to use condition (7.17) with s=7.

When s=3, equation (7.16) has the form

$$f(x_k) + f'(x_k)(x_{k+1} - x_k) + 0.5 f''(x_k)(x_{k+1} - x_k)^2 = 0\,. \tag{7.20}$$

Solving this equation with respect to x_{k+1}, we obtain

$$x_{k+1} = x_k - \frac{f'(x_k)}{f''(x_k)} \pm \frac{|f'(x_k)|}{f''(x_k)}\left(1 - \frac{2f''(x_k)f(x_k)}{\left(f'(x_k)\right)^2}\right)^{1/2}. \tag{7.21}$$

To satisfy Theorem 7.1, we must add the last two terms of equation (7.21) when $f'(x)>0$ and subtract them when $f'(x)<0$. It should also be taken into account that the sum of the last two terms in expression (7.21) approaches zero near the root. This may result in the loss of accuracy. It is known that for quadratic equation $ax^2+bx+c=0$, the following relation is valid:

$$\frac{1}{2a}\left(-b+\sqrt{b^2-4ac}\right)=-\frac{2c}{b+\sqrt{b^2-4ac}}.$$

After applying this conversion to (7.21), we finally arrive at the following iterative scheme:

$$x_{k+1}=x_k-\frac{f(x_k)}{f'(x_k)\left(1+\sqrt{1-\dfrac{2f''(x_k)f(x_k)}{\left(f'(x_k)\right)^2}}\right)},\ k=0,\ 1,\ \ldots. \quad (7.22)$$

An initial value x_0 should be chosen from condition (7.18) with $s=8$.

The polynomial $P_s(x_{k+1})$ may be reduced to a first-degree polynomial without changing the order of the iterative sequence generated. Two possibilities to modify equation (7.20) follow:

1) replace one of the $(x_{k+1}-x_k)$ factors in $(x_{k+1}-x_k)^2$ by $-f(x_k)/f'(x_k)$. After solving the modified equation with respect to $(x_{k+1}-x_k)$ we obtain the following iterative scheme:

$$x_{k+1}=x_k-\frac{f(x_k)f'(x_k)}{\left(f'(x_k)\right)^2-0.5f''(x_k)f(x_k)}. \quad (7.23)$$

$$k=0,\ 1,\ \ldots.$$

This is the Halley method.

2) replace $(x_{k+1}-x_k)^2$ by $(f(x_k)/f'(x_k))^{7\cdot}$ Then after solving the modified equation with respect to $(x_{k+1}-x_k)$ we get

$$x_{k+1}=x_k-\frac{f(x_k)}{f'(x_k)}\left(1-\frac{0.5f''(x_k)f(x_k)}{\left(f'(x_k)\right)^2}\right), \quad (7.24)$$

$$k=0,\ 1,\ \ldots.$$

These methods are more attractive in comparison to iterative scheme (7.22) because they have the same cubic convergence but computation of the square root is avoided.

The following function implements the fixed-point method to find the root of an equation *f(x) = 0* such that the equation could be re-written as *x = g(x)*. It takes three input arguments; the first argument is a function handle. Here, it is the function *g(x)* which is related to function *f(x)*, for which we want to find out the root. The second argument is the initial guess value of root. The last argument is the error tolerance of the search method as discussed previously. The code listing is as follows.

Listing 7.1

```
function y = fixedpoint(g, x0, tol)
% fixedpoint function finds the root of a given function
% f(x) = 0 with the intial guess x0 and accuracy of tol
% such that the equation f(x) = 0 could be written as
```

```
%     x = g(x)
% input:  g - input function such that x = g(x)
%         x0 - initial guess value of root
%         tol - error tolerance
% output: y - root value
y_old = x0;
y_new = feval(g, x0);
i = 0;
while (abs(y_new - y_old) > tol)
    i = i + 1;
    y_old = y_new;
    y_new = feval(g, y_old);
    s = sprintf('Iteration = %d\t\tApproximate root = %5.10f', i, y_new);
    disp(s);
end;
y = y_new;
```

To find the root of an equation $f(x) = cos(x) - x = 0$. We will first re-write the equation as $x = g(x) = cos(x)$. Then, we will implement the following function.

Listing 7.2

```
function y = g(x)
% x = cos(x)
y = cos(x);
```

Then we will call the *fixedpoint* function as follows.

Listing 7.3

```
>> y = fixedpoint(@g, 0, 1e-6)
Iteration = 1      Approximate root = 0.5403023059
....
Iteration = 34     Approximate root = 0.7390855264
y =
0.73908552636192
```

We don't show all the results to save the space. It took 34 iterations to reach the approximate solution within desired error tolerance. We observed that even after trying with different initial root values closer to the root the performance of this method did not improve.

The following function implements the controlled fixed-point iterative method. It is a variation of the fixed-point method. The signature of the function remains almost the same except for the addition of the alpha control parameter. We do not need to transform the equation from $f(x) = 0$ to $x = g(x)$ so the function handle in first argument should point to the m-script implementation of $f(x)$ rather than $g(x)$. The code is listed in the following.

Listing 7.4

```
function y = cfixedpoint(f, x0, alpha, tol)
% cfixedpoint function finds the root of a given function
% f(x) = 0 with the intial guess x0 and accuracy of tol
% such that the iteration is controlled using parameter alpha
% input:  f - input function such that f(x) = 0
%         x0 - initial guess value of root
%         alpha - control parameter
%         tol - error tolerance
% output: y - root value
y_old = x0;
y_new = y_old + alpha * feval(f, x0);
i = 0;
while (abs(y_new - y_old) > tol)
    i = i + 1;
    y_old = y_new;
    y_new = y_old + alpha * feval(f, y_old);
    s = sprintf('Iteration = %d\t\tApproximate root = %5.10f', i, y_new);
    disp(s);
end;
y = y_new
```

We apply this function on the same problem, i.e.,

$f(x) = cos(x) - x = 0$

First, we should implement this function using m-script as given in the following listing.

Listing 7.5

```
function y =f(x)
%f(x) = cos(x) - x = 0
y = cos(x) -x
```

Then, to apply the controlled fixed-point method on this function, we will make the following function call from the interpreter.

Listing 7.6

```
>> y = cfixedpoint(@f, 0, 0.5, 1e-6)
Iteration = 1       Approximate root = 0.7379893799
Iteration = 2       Approximate root = 0.7389060909
Iteration = 3       Approximate root = 0.7390559087
Iteration = 4       Approximate root = 0.7390803638
Iteration = 5       Approximate root = 0.7390843549
Iteration = 6       Approximate root = 0.7390850062
```

---
y =
0.73908500619363

The result obtained is the same as in previous sections but it takes only 6 iterations to solve the same problem. Controller fixed-point method is an enormous improvement in terms of performance compared to the fixed-point method.

7.1.2 Multi-point iterative processes

To construct an iterative process with cubic convergence using one-point iteration, we need to employ the higher-order derivatives of $f(x)$. This may sometimes result in much computational effort. Another way to achieve cubic convergence is to use a multi-point iterative scheme (see (7.5)). One of the procedures to construct such a scheme is based on a superposition. Let us calculate the next approximation x_{k+1} in two stages (note that the first formula represents Newton's method):

$$y_k = x_k - \frac{f(x_k)}{f'(x_k)}, \quad x_{k+1} = y_k - \frac{f(y_k)}{f'(x_k)}.$$

A combination of these two formulae results in the following iterative scheme:

$$x_{k+1} = x_k - \frac{\left(f(x_k) + f\left(x_k - \frac{f(x_k)}{f'(x_k)}\right)\right)}{f'(x_k)}, \tag{7.25}$$

$$k=0, 1, \ldots.$$

$$\beta(x) = x - \frac{f(x)}{f'(x)}, \quad g(x) = \beta(x) - \frac{f(\beta(x))}{f'(x)}.$$

The order of the iterative sequence is three, that is, $e_{k+1}=C(e_k)^3$. In some cases, a multi-point iterative scheme is more efficient in comparison to one-point iteration. Only the function itself and its first derivative are employed and again we have rapid convergence. The construction procedure for iterative scheme (7.25) enables us to adopt the same condition for choosing the initial values as for Newton's method.

By analogy with scheme (7.25), we can construct another iterative scheme:

$$y_k = x_k - \frac{f(x_k)}{f'(x_k)}, \quad x_{k+1} = y_k - \frac{f(x_k)}{f'(y_k)},$$

and after the combination of these expressions we arrive at

$$x_{k+1} = x_k - f(x_k)\left(\frac{1}{f'(x_k)} + \frac{1}{f'\left(x_k - \frac{f(x_k)}{f'(x_k)}\right)}\right),$$

$$k=0, 1, \ldots.$$

This iterative scheme also produces the sequence with the order of convergence $p=3$.

The following function implements Newton's method of root approximation. In this function, the first argument points to the implementation of function *f(x)* whereas the second argument points to the implementation of first derivative of this function, *f'(x)*. The third argument, *x0*, is the initial guess of the root and last argument, *tol*, is the error tolerance.

Listing 7.7

```
function y = newton(f, df, x0, tol)
% newton method finds the root of a given function f(x) = 0
% input:    f - input function such that f(x) = 0
%           df - first derivative of the function f(x)
%           x0 - initial guess value of root
%           tol - error tolerance
% output: y - root value
y_old = x0;
y_new = y_old - feval(f, y_old) / feval(df, y_old);
i = 0;
while (abs(y_new - y_old) > tol)
      i = i + 1;
      y_old = y_new;
      y_new = y_old - feval(f, y_old) / feval(df, y_old);
      s = sprintf('Iteration = %d\t\tApproximate root = %5.10f', i, y_new);
      disp(s);
end;
y = y_new;
```

We use the same example as in the previous sections.

$f(x) = cos(x) - x = 0$

The implementation of this function is shown previously. The derivative of this function is as follows.

$df(x) = -sin(x) - 1$

And the implementation of the derivative is as follows.

Listing 7.8

```
function y = df(x)
y = -sin(x) - 1;
```

Then, to apply *Newton's* method on this function, we will make the following function call from the interpreter.

Listing 7.9

```
>> y = newton(@f, @df, 0, 1e-6)
Iteration = 1     Approximate root = 0.7503638678
Iteration = 2     Approximate root = 0.7391128909
Iteration = 3     Approximate root = 0.7390851334
Iteration = 4     Approximate root = 0.7390851332
```

```
y =
21 0.73908513321516
```

Once again, the result obtained is the same as in previous methods and it takes 4 iterations to reach the solution. This is the best method so far. The only disadvantage of this method is that the user has to write the implementation of the derivative of the function as well. Readers could try to modify the above implementation and use the numerical approach to find out the derivative of the function. Although it will take longer to run the function, the user will not have to write the extra function implementation.

The multi-point method is implemented in the following. The function takes the same arguments as in Newton's method.

Listing 7.10

```
function x = multipoint(f, df, x0, tol)
% newton method finds the root of a given function f(x) = 0
% input:  f - input function such that f(x) = 0
%         df - first derivative of the function f(x)
%         x0 - initial guess value of root
%         tol - error tolerance
% output: x - root value
x_old = x0;
y = x_old - feval(f, x_old) / feval(df, x_old);
x_new = y - feval(f, y) / feval(df, x_old);
i = 0;
while (abs(x_new - x_old) > tol)
   i = i + 1;
   x_old = x_new;
   y = x_old - feval(f, x_old) / feval(df, x_old);
   x_new = y - feval(f, y) / feval(df, x_old);
   s = sprintf('Iteration = %d\t\tApproximate root = %5.10f', i, x_new);
   disp(s);
end;
x = x_new;
```

We use the same example as in the previous sections.

$$f(x) = cos(x) - x = 0$$

$$df(x) = -sin(x) - 1$$

To apply the *multipoint* method, we will make the following function call from the interpreter.

Listing 7.11

```
>> y = multipoint(@f, @df, 0, 1e-6)
Iteration = 1      Approximate root = 0.7379315267
Iteration = 2      Approximate root = 0.7390851331
```

```
Iteration = 3      Approximate root = 0.7390851332
ans =
  0.73908513321516
```

This method proves faster than Newton's method. The result is the same as before and it takes even fewer iterations to reach the solution.

7.1.3 One-point iterative processes with memory

Despite the power of methods with superlinear convergence, their use of the derivatives $f'(x)$ and $f''(x)$ can be troublesome. It often requires much more effort to evaluate $f'(x)$ and $f''(x)$ than $f(x)$. In addition, many applications involve functions that do not have closed-form expressions for their derivatives.

In this section we consider some modifications to one-point iterations that avoid the exact evaluation of derivatives. The first derivative of $f(x)$ at the point x_k may be approximately evaluated in the following manner:

$$f'(x_k) \approx \frac{f(x_k) - f(x_{k-1})}{x_k - x_{k-1}}.$$

Substituting this expression in (7.19) we obtain

$$x_{k+1} = x_k - \frac{f(x_k)(x_k - x_{k-1})}{f(x_k) - f(x_{k-1})}, \; k=1, 2, \ldots. \tag{7.26}$$

This is the secant method. Convergence of this iterative scheme is described by expression (7.8) with some constant C and $p \approx 1.62$, which is slower than Newton's method ($p=2$). However, the secant method requires just one evaluation of $f(x_k)$ per iteration (we can use the value $f(x_{k-1})$ from the previous iteration). To start an iterative process we need to choose two initial values x_0 and x_1. To define these initial values, we can use the same condition we used for Newton's method, except that the condition should be applied to x_0 and x_1:

$$f(x_m)f''(x_m) > 0, \; m=0, 1. \tag{7.27}$$

The implementation of secant method in the following requires two guess values of the root, unlike the previous method.

Listing 7.12

```
function y = secant(f, df, x0, x1, tol)
% This method finds the root of a given function f(x) = 0
% input:   f - input function such that f(x) = 0
%          x0 – first initial guess value of root
%          x1 - second initial guess value of root
%          tol - error tolerance
% output: y - root value
y_old_old = x0;
y_old = x1;
y_new = y_old - feval(f, y_old) *
                (y_old - y_old_old)/ (feval(f, y_old) - feval(f, y_old_old));
```

```
i = 0;
while (abs(y_new - y_old) > tol)
      i = i + 1;
      y_old_old = y_old;
      y_old = y_new;
      y_new = y_old - feval(f, y_old) *
              (y_old - y_old_old)/ (feval(f, y_old) - feval(f, y_old_old));
      s = sprintf('Iteration = %d\t\tApproximate root = %5.10f', i, y_new);
      disp(s);
end;
y = y_new;
```

We use the same example as in the previous sections.

$f(x) = cos(x) - x$

$df(x) = -sin(x) - 1$

The implementation of these functions is shown previously. To apply the *secant* method, we will make the following function call from the interpreter.

Listing 7.13

```
>> y = secant(@f, @df, 0, 1, 1e-6)
Iteration = 1      Approximate root = 0.7362989976
Iteration = 2      Approximate root = 0.7391193619
Iteration = 3      Approximate root = 0.7390851121
Iteration = 4      Approximate root = 0.7390851332
y =
0.73908513321500
```

Secant method took only 4 iterations to find the root. Its performance is comparable to the multi-point method to solve this example.

7.2 Exercises

Find the root of equation $f(x)=0$. If the equation has more than one root, then find the root that has the smallest magnitude. Calculate the root with prescribed tolerance using one of the iterative methods. To solve the problem find an interval that contains the root and initial guess that initiates the convergent iterative process. Suggested iterative methods are

(1) simple iteration with Aitken's δ^2-process,
(2) Newton's method (7.19),
(3) one-point iteration (7.22),
(4) one-point iteration (7.23),
(5) one-point iteration (7.24),
(6) multi-point iteration (7.25),
(7) secant method (7.26)

Set of equations to solve:

1. $\ln(x) + (x+1)^3 = 0$
2. $x \cdot 2^x - 1 = 0$
3. $\sqrt{x+1} - \frac{1}{x} = 0$
4. $x - \cos x = 0$
5. $3x + \cos x + 1 = 0$
6. $x + \lg(x) - 0.5 = 0$
7. $2 - x - \ln(x) = 0$
8. $(x-1)^2 - 0.5e^x = 0$
9. $(2-x)e^x = 0$
10. $2.2x - 2^x = 0$
11. $x^2 + 4\sin x = 0$
12. $2x - \lg(x) - 7 = 0$
13. $5x - 8\ln(x) - 8 = 0$
14. $3x - e^x = 0$
15. $x(x+1)^2 - 1 = 0$
16. $x - (x+1)^3 = 0$
17. $x^2 - \sin(x) = 0$
18. $x^3 - \sin(x) = 0$
19. $x - \sqrt{\ln(x+2)} = 0$
20. $x^2 - \ln(x+1) = 0$
21. $2x + \ln(x) + 0.5 = 0$
22. $2x + \cos(x) - 0.5 = 0$
23. $\sin(0.5x) + 1 - x^2 = 0 \,; x > 0$
24. $0.5x + \ln(x-1) - 0.5 = 0$
25. $\sin(0.5 + x) - 2x + 0.5 = 0$
26. $\ln(2+x) + 2x - 3 = 0$
27. $\ln(1+2x) - 2 + x = 0$

28. $2\sin(x - 0.6) - 1.5 + x = 0$
29. $x + \ln(1 + x) - 1.5 = 0$
30. $x + \cos(x) - 1 = 0$

8

Solving Systems of Linear Equations

8.1 Linear Algebra Background

To begin with, we will consider some basic definitions from linear algebra needed for understanding of matrix computations. We will restrict our consideration to the case when matrices and vectors are real. Vectors, then, are objects of the N-dimensional Euclidean space R^N. A list of the operations defined in R^N follows.

- Addition:

$$\mathbf{x} + \mathbf{y} = \mathbf{y} + \mathbf{x}, \; (\mathbf{x} + \mathbf{y}) + \mathbf{z} = \mathbf{x} + (\mathbf{y} + \mathbf{z}).$$

- Multiplication by scalar:

$$\alpha(\mathbf{x} + \mathbf{y}) = \alpha\mathbf{x} + \alpha\mathbf{y}.$$

In addition, there is correspondence between any pair of vectors $\mathbf{x}, \mathbf{y} \in R^N$ and a real scalar $(\mathbf{x}, \mathbf{y})$ called the dot product. The following axioms for the dot product are postulated:

$$(\mathbf{x}, \mathbf{y}) = (\mathbf{y}, \mathbf{x}),$$
$$(\alpha\mathbf{x}, \mathbf{y}) = \alpha(\mathbf{x}, \mathbf{y}), \; \alpha \text{ is a real scalar},$$
$$(\mathbf{x} + \mathbf{y}, \mathbf{z}) = (\mathbf{x}, \mathbf{z}) + (\mathbf{y}, \mathbf{z}),$$
$$(\mathbf{x}, \mathbf{x}) > 0 \text{ for } \mathbf{x} \neq \mathbf{0}, \text{ here } \mathbf{0} \text{ is the zero vector, } (\mathbf{0}, \mathbf{0}) = 0.$$

Column vectors:

$$\mathbf{b}_1 = \begin{pmatrix} 1 \\ 0 \\ \cdot \\ \cdot \\ \cdot \\ 0 \end{pmatrix}, \mathbf{b}_2 = \begin{pmatrix} 0 \\ 1 \\ \cdot \\ \cdot \\ \cdot \\ 0 \end{pmatrix}, \ldots, \mathbf{b}_N = \begin{pmatrix} 0 \\ \cdot \\ \cdot \\ \cdot \\ 0 \\ 1 \end{pmatrix}$$

form a standard basis of R^N. All other vectors are expressed in terms of the basis $\mathbf{b}_n$ (n=1, ..., N):

$$\mathbf{x} = x_1\mathbf{b}_1 + \ldots + x_N\mathbf{b}_N .$$

The coefficients x_n are the components of a vector $\mathbf{x}$ and this representation is commonly written in a compact form $\mathbf{x}=(x_1, \ldots, x_N)^{\mathrm{T}}$. The superscript T indicates the transposition that changes column vectors into row vectors and vice versa. The dot product of two vectors $\mathbf{x}=(x_1, \ldots, x_N)^{\mathrm{T}}$ and $\mathbf{y}=(y_1, \ldots, y_N)^{\mathrm{T}}$ is

$$(\mathbf{x}, \mathbf{y}) = \mathbf{x}^{\mathrm{T}}\mathbf{y} = \sum_{n=1}^{N} x_n y_n .$$

Two vectors $\mathbf{x}$ and $\mathbf{y}$ are orthogonal if $(\mathbf{x}, \mathbf{y})=0$.

To work with an object, we need some way to measure it. To do so, a vector norm is introduced. A norm of a vector $\mathbf{x}$ is a real scalar $\|\mathbf{x}\|$ with the following properties:

$$\|\mathbf{x}\| > 0 \text{ for } \mathbf{x}\neq\mathbf{0} \text{ and } \|\mathbf{0}\| = 0,$$

$$\|\alpha\mathbf{x}\| = |\alpha| \cdot \|\mathbf{x}\|, \alpha \text{ is a real scalar},$$

$$\|\mathbf{x} + \mathbf{y}\| \leq \|\mathbf{x}\| + \|\mathbf{y}\| \text{ for any } \mathbf{x}, \mathbf{y}.$$

For a vector $\mathbf{x}=(x_1, \ldots, x_N)^{\mathrm{T}}$, scalar

$$\|\mathbf{x}\|_p = \left(\sum_{n=1}^{N} |x_n|^p\right)^{1/p}$$

is a norm for any $p\geq1$ (the Hölder p-norm). The following norms are commonly used:

$$\|\mathbf{x}\|_1 = \sum_{n=1}^{N} |x_n|,$$

$$\|\mathbf{x}\|_2 = \left(\sum_{n=1}^{N} x_n^2\right)^{1/2} = \sqrt{(\mathbf{x},\mathbf{x})},$$

$$\|\mathbf{x}\|_\infty = \max_n |x_n|.$$

Every square matrix generates a linear operator in the Euclidean space R^N. If $\mathbf{y}$ is a linear function of $\mathbf{x}$, then

$$\mathbf{y} = \mathbf{A}\mathbf{x}\,,\ \mathbf{A} = \{a_{nm}\}$$

or

$$y_n = \sum_{m=1}^{N} a_{nm} x_m\ ,\ n{=}1,\ \ldots,\ N.$$

The rules for manipulating matrices are those required by the linear mappings. Thus $\mathbf{C}=\mathbf{A}+\mathbf{B}$ is the sum of two linear functions represented by **A** and **B**, where $c_{nm}=a_{nm}+b_{nm}$. The product **AB** represents the effect of applying the mapping **B**, then applying the mapping **A**. If we have $\mathbf{y}=\mathbf{A}(\mathbf{B}\mathbf{x})=\mathbf{C}\mathbf{x}$, then components of matrix **C** are given by the formula

$$c_{nm} = \sum_{k=1}^{N} a_{nk} b_{km}\ ,\ n,\ m{=}1,\ \ldots,\ N.$$

The transposition of **A** is obtained by reflecting **A** about its main diagonal:

$$\mathbf{A}^{\mathrm{T}} = \{a_{nm}\}^{\mathrm{T}} = \{a_{mn}\}\,.$$

A diagonal matrix **D** is all zeros except for d_n on the main diagonal: $\mathbf{D}=\mathrm{diag}(d_1,\ \ldots,\ d_N)$.

Now, let us consider norms of a matrix. Scalar $||\mathbf{A}||$ is a norm of a matrix **A** if the following conditions are satisfied:

$$\|\mathbf{A}\| > 0 \text{ for } \mathbf{A}\neq\mathbf{O}\text{, here } \mathbf{O} \text{ is the zero matrix and } ||\mathbf{O}||=0,$$
$$\|\alpha\mathbf{A}\| = |\alpha|\cdot\|\mathbf{A}\|\,,\ \alpha \text{ is a real scalar,}$$
$$\|\mathbf{A}+\mathbf{B}\| \le \|\mathbf{A}\| + \|\mathbf{B}\|\,,$$
$$\|\mathbf{A}\mathbf{B}\| \le \|\mathbf{A}\|\cdot\|\mathbf{B}\|\,.$$

The following norms are commonly used:

$$\|\mathbf{A}\|_1 = \max_m \sum_{n=1}^{N} |a_{nm}|\,,$$

$$\|\mathbf{A}\|_2 = \left(\sum_{n=1}^{N}\sum_{m=1}^{N} a_{nm}^2\right)^{1/2},$$

$$\|\mathbf{A}\|_\infty = \max_n \sum_{m=1}^{N} |a_{nm}|\,.$$

In what follows the notation $||\cdot||$ will denote some norm given above. A matrix norm is consistent with a vector norm if $||\mathbf{A}\mathbf{x}||\le||\mathbf{A}||\cdot||\mathbf{x}||$. For example, $||\mathbf{A}||_1$ is consistent with $||\mathbf{x}||_1$, $||\mathbf{A}||_2$ with $||\mathbf{x}||_2$, and $||\mathbf{A}||_\infty$ with $||\mathbf{x}||_\infty$.

The property of matrix definiteness plays an important role in matrix computations. So, a matrix **A** is positive (negative) definite if $(\mathbf{x},\ \mathbf{A})>0$ (<0) for all nonzero vectors **x**. If the sign of $(\mathbf{x},\ \mathbf{A})$ depends on vector x, then matrix **A** is called indefinite.

In the following, we demonstrate how to calculate the dot product of two vectors.

Listing 8.1

```
>> a = [1 2 3];
>> b = [-1 0 4];
>> c = dot(a, b)
c =
11.0000
>> d = sum(a.*b)
d =
11.0000
```

You could either use the *dot* function to compute the dot product of two vectors or you could use the dot product operator ".*" and then apply sum to get the dot product. The latter approach is better in performance.

The *norm* function, as demonstrated below, could be used to find out different norm values of a vector. The second argument determines the order of norm value. In the absence of second argument, the function returns 2-norm. This function also works for a matrix argument in the similar manner.

Listing 8.2

```
>> a = [1 2 3];
>> norm(a)
ans =
3.7417
>> norm(a, 1)
ans =
6.0000
>> norm(a, 2)
ans =
3.7417
>> norm(a, Inf)
ans =
3.0000
```

8.2 Systems of Linear Equations

A system of linear equations may be represented as

$$\mathbf{Ax} = \mathbf{f}\,,\ \det(\mathbf{A})\neq 0 \tag{8.1}$$

or, in expanded form,

$$a_{11}x_1 + a_{12}x_2 + \ldots + a_{1N}x_N = f_1\,,$$

$$\ldots,$$

$$a_{N1}x_1 + a_{N2}x_2 + \ldots + a_{NN}x_N = f_N\,,$$

where $\mathbf{A}=\{a_{nm}\}$ is a square matrix of coefficients of size N by N, $\mathbf{x}=(x_1, \ldots, x_N)^T$ is a vector of unknowns of size N, and $\mathbf{f}=(f_1, \ldots, f_N)^T$ is a known right-hand vector of size N.

There are numerous physical systems which naturally lead to linear equations, for example, the static analysis of structures (bridges, buildings, and aircraft frames) and the design of electrical networks. Many physical systems are modeled by differential equations, which often cannot be solved analytically. As a result, the finite difference approach has to be used to approximate differential equations. This often results in a system of linear equations (if the differential equation is linear) for approximate values of the solution. Also, systems of linear equations may arise as an intermediate stage in other computational processes. Matrix inversion is also an inefficient approach (if we have $\mathbf{A}^{-1}$, then computation of $\mathbf{x}$ is easy: $\mathbf{x}=\mathbf{A}^{-1}\mathbf{f}$), because computing an inverse matrix requires about three times as much computation and twice as much storage as does solving a system of linear equations.

To adopt one or another method for the solution of system (8.1) we should take into account properties of matrix **A**, such as symmetry, definiteness, band structure, and sparsity. This allows us to solve a system of linear equations more efficiently. The adoption of a method also depends on the type of problem under consideration, i.e., whether the problem requires solving a single linear system $\mathbf{Ax}=\mathbf{f}$ or solving a set of linear equations $\mathbf{Ax}_k=\mathbf{f}_k$, $k=1, \ldots, M$, where matrix **A** is the same for all k.

The following code listing demonstrates the methods to calculate the inverse and determinate of a matrix. There are two ways to calculate the inverse of a matrix. You could either use the *inv* function or you could take the reciprocal of the matrix.

Listing 8.3

```
>> u = rand(3);
u =
0.9418   0.0777   0.1260
0.2466   0.1872   0.5753
0.9774   0.5600   0.0026
>> inv(u)
ans =
1.2140   -0.2655   -0.0797
-2.1196   0.4555   1.9274
0.1694   1.7039   -0.5932
>> 1/u
ans =
1.2140   -0.2655   -0.0797
-2.1196   0.4555   1.9274
0.1694   1.7039   -0.5932
>> det(u)
ans =
-0.2650
```

8.3 Types of Matrices that arise from Applications and Analysis

8.3.1 Sparse matrices

Matrices with most of the elements equal to zero are called sparse. We can define a matrix **A** of size N by N as sparse if the number of its nonzero elements $\sim N^{1+\gamma}$ ($\gamma \leq 0.5$). For example, for $N=10^3$ and $\gamma=0.5$, the number of nonzero elements is 31622 (the total number of elements is 10^6).

8.3.2 Band matrices

Many applications lead to matrices that are not only sparse, but have a band pattern in their nonzero elements. Matrix $\mathbf{A}=\{a_{nm}\}$ is called a band matrix if $a_{nm}=0$ for $|n-m|>k$, where $l=2k+1$ is the band width:

$$\mathbf{A} = \begin{pmatrix} * & * & & & & \\ * & \ldots & * & 0 & & \\ & * & \leftarrow l \rightarrow & * & & \\ & 0 & * & \ldots & * \\ & & & * & * \end{pmatrix}.$$

8.3.3 Symmetric positive definite matrices

Many physical problems lead to symmetric positive definite (SPD) matrices. The properties of SPD matrices follow:

- symmetry:

$$\mathbf{A} = \mathbf{A}^{\mathrm{T}}.$$

- positive definiteness:

$$(\mathbf{x}, \mathbf{A}\mathbf{x}) > 0 \text{ for all nonzero vectors } \mathbf{x}.$$

It is rather difficult to determine whether a matrix is positive definite. Fortunately, there are more easily verified criteria for identifying members of this important class. First, all eigenvalues of a positive definite matrix are positive. This can be easily verified with the aid of the Gerschgorin theorem (see Chapter 4). Second, if **A** is an N by N positive definite matrix, then it has the following properties: (1) $a_{nm}>0$ for all n; (2) $a_{nn}a_{mm}>(a_{nm})^2$ for $n \neq m$; and (3) the element of the largest modulus must lie on the leading diagonal.

8.3.4 Triangular matrices

Two types of triangular matrices follow, both of which are easy to work with.

- Lower triangular matrix:

$$\mathbf{L} = \begin{pmatrix} * & & & & \\ * & * & & 0 & \\ * & . & * & & \\ * & . & . & * & \\ * & * & * & * & * \end{pmatrix}, \ a_{nm}=0 \text{ for } m>n,$$

- Upper triangular matrix:

$$\mathbf{U} = \begin{pmatrix} * & * & * & * & * \\ & * & . & . & * \\ & & * & . & * \\ & 0 & & * & * \\ & & & & * \end{pmatrix}, \ a_{nm}=0 \text{ for } m<n.$$

The following function implements the LU decomposition of a tri-diagonal matrix. This utility function finds its application in a number of physical problems.

Listing 8.4

```
function [L, U] = ludecomp(A)
% ludecomp function decompose a matrix into
% Lower matrix (L) and upper matrix (U)
% input:   A - input square matrix
% output: L - lower triangular factor
%            U - upper triangular factor
n = length(A);
L = eye(n);
U = zeros(n);
U(1,1) = A(1,1);
for i=2:n
      L(i, i-1) = A(i, i-1)/A(i-1, i-1);
      U(i-1, i) = A(i-1, i);
      U(i, i) = A(i, i) - L(i, i-1) * A(i-1, i);
      A(i, i) = U(i, i);
end;
```

The application of this function is demonstrated in the following listing.

Listing 8.5

```
>> A = [2 -2 0 0 0; -2 5 -6 0 0; 0 -6 16 12 0; 0 0 12 39 -6; 0 0 0 -6 14];
>> [L U] = ludecomp(A)
L =
```

```
    1.0000   0.0000   0.0000   0.0000   0.0000
   -1.0000   1.0000   0.0000   0.0000   0.0000
    0.0000  -2.0000   1.0000   0.0000   0.0000
    0.0000   0.0000   3.0000   1.0000   0.0000
    0.0000   0.0000   0.0000  -2.0000   1.0000
U =
    2.0000  -2.0000   0.0000   0.0000   0.0000
    0.0000   3.0000  -6.0000   0.0000   0.0000
    0.0000   0.0000   4.0000  12.0000   0.0000
    0.0000   0.0000   0.0000   3.0000  -6.0000
    0.0000   0.0000   0.0000   0.0000   2.0000
>> A = L * U
A =
    2.0000  -2.0000   0.0000   0.0000   0.0000
   -2.0000   5.0000  -6.0000   0.0000   0.0000
    0.0000  -6.0000  16.0000  12.0000   0.0000
    0.0000   0.0000  12.0000  39.0000  -6.0000
    0.0000   0.0000   0.0000  -6.0000  14.0000
```

A system of linear equations **Lx=f** can be solved by forward substitution:

$$x_1 = f_1 / a_{11},$$

$$x_n = \frac{1}{a_{nn}}\left(f_n - \sum_{m=1}^{n-1} a_{nm} x_m \right), \; n=2, \ldots, N. \tag{8.2}$$

In an analogous way, a system of linear equations **Ux=f** can be solved by backward substitution:

$$x_N = f_N / a_{NN},$$

$$x_n = \frac{1}{a_{nn}}\left(f_n - \sum_{m=n+1}^{N} a_{nm} x_m \right), \; n=N-1, \ldots, 1. \tag{8.3}$$

The following implementation of forward substitution method is used to solve a system of equations when the coefficient matrix is a lower triangular matrix. The function takes two arguments; the lower triangular coefficient matrix and the right-hand side vector. The output vector is the solution of the systems of equation.

Listing 8.6

```
function x = fsubstt(L, f)
%fsubstt function solves the linear system of equations
% using forward substitution method Lx = f such that L
% is the lower triangular matrix
% input:   L - lower triangular matrix
%          f - right-hand side vector
% output: x - solution vector
n = length(f);
x = zeros(n, 1);
```

```
x(1) = f(1) / L(1, 1);
for i = 2 : n
        x(i) = (f(i) - L(i, 1:i-1) * x(1:i-1)) / L(i, i);
end;
```

Say, we have the following system of equations given in a matrix form.

$$\begin{pmatrix} 16 & 0 & 0 & 0 \\ 5 & 11 & 0 & 0 \\ 9 & 7 & 6 & 0 \\ 4 & 14 & 15 & 1 \end{pmatrix}\begin{pmatrix} x_1 \\ x_2 \\ x_3 \\ x_4 \end{pmatrix} = \begin{pmatrix} 16 \\ 27 \\ 41 \\ 81 \end{pmatrix}$$

Since the coefficient matrix is a lower triangular matrix, forward substitution method could be applied to solve the problem, as shown in the following.

Listing 8.7

```
>> L = [16 0 0 0; 5 11 0 0; 9 7 6 0; 4 14 15 1];
>> f = [16; 27; 41; 81]
>> x = fsubstt(L, f)
x =
1.0000
2.0000
3.0000
4.0000
```

The following implementation of backward substitution method is used to solve a system of equations when the coefficient matrix is an upper triangular matrix. The function takes two arguments; the upper triangular coefficient matrix and the right-hand side vector. The output vector is the solution of the systems of equation.

Listing 8.8

```
function x = bsubstt(U, f)
% bsubstt function solves the linear system of equations
% using backward substitution method Ux = f such that
%U is the upper triangular matrix
% input:   U - upper triangular matrix
%            f - right-hand side vector
% output:  x - solution vector
n = length(f);
x = zeros(n, 1);
x(n) = f(n) / U(n, n);
for i = n-1 : -1 : 1
        x(i) = (f(i) - U(i, i+1:n) * x(i+1:n)) / U(i, i);
end;
```

Say, we have the following system of equations given in a matrix form.

$$\begin{pmatrix} 16 & 2 & 3 & 13 \\ 0 & 11 & 10 & 8 \\ 0 & 0 & 6 & 12 \\ 0 & 0 & 0 & 1 \end{pmatrix} \begin{pmatrix} x_1 \\ x_2 \\ x_3 \\ x_4 \end{pmatrix} = \begin{pmatrix} 81 \\ 53 \\ 60 \\ 3 \end{pmatrix}$$

Since the coefficient matrix is an upper triangular matrix, backward substitution method could be applied to solve the problem, as shown in the following.

Listing 8.7

```
>> U = [16 2 3 13; 0 11 10 8; 0 0 6 12; 0 0 0 1];
>> f = [81; 53; 60; 3];
>> x = bsubstt(U, f)
x =
2.0000
-1.0000
4.0000
3.0000
```

8.3.5 Orthogonal matrices

If **A** is an orthogonal matrix, then $\mathbf{A}^T\mathbf{A}=\mathbf{I}$ or $\mathbf{A}^T=\mathbf{A}^{-1}$, and it is easy to solve linear system (8.1) as $\mathbf{x}=\mathbf{A}^T\mathbf{f}$.

8.3.6 Reducible matrices

A square matrix **A** is reducible if its rows and columns can be permuted to bring it into the form:

$$\mathbf{A} = \begin{pmatrix} \mathbf{B} & \mathbf{C} \\ \mathbf{O} & \mathbf{D} \end{pmatrix},$$

where **B** is a square matrix of size M by M, **D** is a square matrix of size L by L, **C** is a matrix of size M by L, and **O** is zero matrix of size L by M ($M+L=N$). This property allows us to break the problem into two smaller ones, usually with a very substantial saving in work. Let us split each of the vectors **x** and **f** into two vectors:

$$\mathbf{x}_1 = (x_1,\ldots,x_M)^T \text{ and } \mathbf{x}_2 = (x_{M+1},\ldots,x_N)^T,$$

$$\mathbf{f}_1 = (f_1,\ldots,f_M)^T \text{ and } \mathbf{f}_2 = (f_{M+1},\ldots,f_N)^T.$$

Then system $\mathbf{Ax}=\mathbf{f}$ can be solved in two stages, first

$$\mathbf{D}\mathbf{x}_2 = \mathbf{f}_2,$$

then

$$\mathbf{B}\mathbf{x}_1 = \mathbf{f}_1 - \mathbf{C}\mathbf{x}_2.$$

8.3.7 Matrices with diagonal dominance

If **A** is a matrix with strong diagonal dominance, then

$$|a_{nn}| > \sum_{\substack{m=1 \\ m \neq n}}^{N} |a_{nm}| = sum_n\,,\ n=1,\ \ldots,\ N.$$

If **A** is a matrix with weak diagonal dominance, then

$$|a_{nn}| \geq sum_n\,,\ n=1,\ \ldots,\ N,$$

and at least one of the inequalities is satisfied as a strict one.

8.4 Error Sources

When we solve problem (8.1) by some method, three types of error may occur. First of all, we assume that problem (8.1) is posed exactly. Thus we neglect inevitable errors connected with formulating a problem as a model of scientific process (measurement errors, insufficiently precise basic hypotheses, etc.).

A vector **f** in (8.1) is not given a priori but must be computed by the data of the exact problem. These computations are performed with a certain error, and instead of **f** we actually obtain some "perturbed" vector $\mathbf{f}+\delta\mathbf{f}$. Then instead of problem (8.1) we obtain the "perturbed" problem

$$\mathbf{A}(\mathbf{x} + \delta\mathbf{x}) = \mathbf{A}\mathbf{y} = \mathbf{f} + \delta\mathbf{f}\,. \tag{8.4}$$

We assume this problem is also solvable, and y is the solution. Then

$$e_p = \|\mathbf{y} - \mathbf{x}^*\|$$

is the perturbation error.

Now the perturbed system (8.4) must be solved. Let us assume that we have some algorithm which produces after a finite number of operations either an exact vector **y** (if operations are performed error free) or some vector **z** such, that the error

$$e_a = \|\mathbf{z} - \mathbf{y}\|$$

can be estimated and the error bound can be made arbitrarily small. The quantity e_a is called the algorithm error.

In fact any algorithm contains associated floating-point operations, which are connected with errors. These errors amount to the situation that instead of **z** we obtain some other vector **w** and the quantity

$$e_r = \|\mathbf{w} - \mathbf{z}\|$$

is called the rounding error.

Finally, instead of the wanted solution $\mathbf{x}^*$ we compute some vector **w**. It is natural to take this vector as the approximate solution to problem (8.1) and its error is

$$e = \|\mathbf{w} - \mathbf{x}^*\| \leq e_p + e_a + e_r\,.$$

8.5 Condition Number

As we discussed before, we have an original problem $\mathbf{Ax}=\mathbf{f}$ and a perturbed problem $\mathbf{A}(\mathbf{x}+\delta\mathbf{x})=\mathbf{f}+\delta\mathbf{f}$, where $\delta\mathbf{f}$ represents the perturbation in both the right-hand side and **A**. The objective is to estimate how the uncertainty $\delta\mathbf{f}$ in **f** is transmitted to $\delta\mathbf{x}$, that is, to estimate the quantity

$$\frac{\|\delta\mathbf{x}\|/\|\mathbf{x}\|}{\|\delta\mathbf{f}\|/\|\mathbf{f}\|}. \tag{8.5}$$

It follows from (8.1) and (8.4) that

$$||\mathbf{f}||=||\mathbf{Ax}||\leq||\mathbf{A}||\cdot||\mathbf{x}||$$

and

$$||\delta\mathbf{x}||=||\mathbf{A}^{-1}\delta\mathbf{f}|| \leq ||\mathbf{A}^{-1}||\cdot||\delta\mathbf{f}||.$$

If we substitute this into the previous expression, we get

$$\frac{\|\delta\mathbf{x}\|/\|\mathbf{x}\|}{\|\delta\mathbf{f}\|/\|\mathbf{f}\|} \leq \|\mathbf{A}\|\cdot\|\mathbf{A}^{-1}\|$$

or

$$\frac{\|\delta\mathbf{x}\|}{\|\mathbf{x}\|} \leq \|\mathbf{A}\|\cdot\|\mathbf{A}^{-1}\|\frac{\|\delta\mathbf{f}\|}{\|\mathbf{f}\|}.$$

The quantity

$$\mathrm{cond}(\mathbf{A}) = \|\mathbf{A}\|\cdot\|\mathbf{A}^{-1}\|$$

is called the standard condition number. This number estimates the maximum magnification of the uncertainty $\delta\mathbf{f}$. If $\mathrm{cond}(\mathbf{A})$ is relatively small, then system (8.1) is well conditioned and $\delta\mathbf{x}$ is relatively small as well. For example, if $||\delta\mathbf{f}||/||\mathbf{f}|| = 10^{-14}$ and we want $||\delta\mathbf{x}||/||\mathbf{x}||=10^{-6}$, then $\mathrm{cond}(\mathbf{A})=10^{8}$ is quite acceptable. If $\mathrm{cond}(\mathbf{A})$ is relatively large, then we face an ill-conditioned system and we should use some special procedures to obtain an acceptable result.

The direct calculation of the condition number is usually inefficient, as we need to compute an inverse matrix. Therefore, we must resort to estimating the condition number. numEclipse provides the *cond* function to estimate the condition number as demonstrated in the following code listing.

Listing 8.8

```
>> u = rand(3);
u =
0.9418  0.0777  0.1260
0.2466  0.1872  0.5753
0.9774  0.5600  0.0026
>> cond(u)
ans =
4.5682
```

8.6 Direct Methods

In this part we will consider direct methods. The distinctive feature of these methods is that, ignoring round-off errors, they produce the true solution after a finite number of steps. Let us discuss the basic principles of these methods.

8.6.1 Basic principles of direct methods

The solution is obtained after first factorizing **A** as a product of triangular matrices, and the form of factorization depends on properties of a matrix **A**. For general matrices we obtain

$$\mathbf{A}=\mathbf{LU},$$

where **L** is a lower triangular matrix with unit diagonal elements, and **U** is an upper triangular one. For SPD matrices we use the Cholesky factorization

$$\mathbf{A} = \mathbf{LL}^{\mathrm{T}},$$

where **L** is a lower triangular matrix. Hence, a system **Ax**=**f** can be transformed into two systems of equations:

$$\mathbf{Ly}=\mathbf{f},\ \mathbf{Ux}=\mathbf{y} \quad \text{or} \quad \mathbf{Ly}=\mathbf{f},\ \mathbf{L}^{\mathrm{T}}\mathbf{x}=\mathbf{y}. \tag{8.6}$$

The systems are solved simultaneously by forward and backward substitutions, respectively (see (8.2) and (8.3)). The error sources for these methods are the perturbation errors in factors **L** and **U** and the rounding errors in solving systems (8.6). An estimation of the work for the LU-factorization yields $N_{\mathrm{ops}}\sim N^3$, and the Cholesky factorization requires half as much work and storage as the method of LU-factorization.

There is another factorization that is useful. Find an orthogonal matrix **Q** and upper triangular matrix **R** so that **QAx**=**Rx**=**Qf**; therefore, **x** may be obtained by backward substitution on the vector **Qf**. The attraction of the orthogonal factorization lies in the fact that multiplying **A** or **f** by **Q** does not magnify round-off errors or uncertainties associated with the matrix **A** or vector **f**. To see this, note that

$$(\mathbf{Qx},\mathbf{Qx}) = (\mathbf{Qx})^{\mathrm{T}}(\mathbf{Qx}) = \mathbf{x}^{\mathrm{T}}\mathbf{Q}^{\mathrm{T}}\mathbf{Qx} = \mathbf{x}^{\mathrm{T}}\mathbf{x},$$

since $\mathbf{Q}^{\mathrm{T}}\mathbf{Q}=\mathbf{I}$. So the advantage of this approach is that it produces a numerically stable method. However, this method requires twice as much work as the method of LU-factorization. That is why the application of orthogonal transformations to the solution of systems of linear equations is limited to some special cases. As we will see, these transformations have also found application in the eigenvalue problem.

Now we consider the sweep method. This technique is based on a special transformation that can be applied to systems with a band matrix. We restrict ourselves to the case where the matrix of system (8.1) is a tridiagonal one. It is convenient for us to represent a system of equations with a tridiagonal matrix in the explicit form:

$$\begin{aligned} b_1 x_1 + c_1 x_2 &= f_1, \\ a_n x_{n-1} + b_n x_n + c_n x_{n+1} &= f_n,\ n=2,\ \ldots,\ N-1, \\ a_N x_{N-1} + b_N x_N &= f_N. \end{aligned} \tag{8.7}$$

The basic idea of the sweep method is to represent a solution in the following form:

$$x_n = \alpha_n x_{n+1} + \beta_n, \tag{8.8}$$

which is at least possible for the first equation. Let coefficients α_{n-1} and β_{n-1} be known, then we can write

$$x_{n-1} = \alpha_{n-1}x_n + \beta_{n-1},$$
$$a_n x_{n-1} + b_n x_n + c_n x_{n+1} = f_n.$$

After eliminating x_{n-1}, we rewrite the previous expression as follows:

$$x_n = -\frac{c_n}{a_n\alpha_{n-1} + b_n}x_{n+1} + \frac{f_n - a_n\beta_{n-1}}{a_n\alpha_{n-1} + b_n}.$$

Therefore, coefficients α_n and β_n are calculated according to the relations

$$\alpha_n = -\frac{c_n}{a_n\alpha_{n-1} + b_n},$$
$$\beta_n = \frac{f_n - a_n\beta_{n-1}}{a_n\alpha_{n-1} + b_n},$$
$$\alpha_1 = -\frac{c_1}{b_1},\ \beta_1 = \frac{f_1}{b_1},$$

where n=2, …, N.

Once the parameters α_n and β_n have been computed, we can easily solve system (8.7). To begin with, let us write expression (8.8) for $n=N-1$ and the last equation (8.7):

$$x_{N-1} = \alpha_{N-1}x_N + \beta_{N-1},$$
$$a_N x_{N-1} + b_N x_N = f_N.$$

Eliminating x_{N-1}, we arrive at

$$x_N = \frac{f_N - a_N\beta_{N-1}}{a_N\alpha_{N-1} + b_N}.$$

All other components x_n are calculated using (8.10) for $n=N-1$, …, 1. The sweep method is a very efficient technique for solving systems of equations with a tridiagonal matrix because $N_{\text{ops}}=8N-6$.

The following theorem describes conditions which provide stability of the sweep method.

Theorem 8.1

Assume that the coefficients of system (8.7) satisfy the conditions

$$b_1 \neq 0,\ b_N \neq 0,\ a_n \neq 0,\ c_n \neq 0,\ n=2,\ \ldots,\ N-1, \qquad (8.9)$$
$$|b_n| \geq |a_n| + |c_n|,\ |b_1| \geq |c_1|,\ |b_N| \geq |a_N|,$$

where at least one of the inequalities is satisfied as a strict one. Then the following inequalities hold true:

$$|\alpha_n| \leq 1$$

and

$$|a_n\alpha_{n-1} + b_n| \geq \min_n |c_n|,$$

where n=2, …, N. Thus, conditions (8.9) turn out to be sufficient for performing the sweep method.

The following function implements the direct method of solving a system of linear equations such that the coefficient matrix is factored into lower- and upper-triangular matrices. The function takes these matrices as arguments along with the right-hand side vector. The implementation is simple and based on the last two implementations of forward and backward substitution. In the first step, forward substitution is applied and then the backward substitution is carried out to find the solution.

Listing 8.9

```
function x = ludirect(L, U, f)
% ludirect function solves the linear system of equations
% when the L & U factors are available Ly = f, Ux = y
% input:   L - lower triangular factor
%          U - upper triangular factor
%          f - righthand side vector
% output: x - solution vector
y = fsubstt(L, f);
x = bsubstt(U, y);
```

We have the following system of equations given in a matrix form.

$$\begin{pmatrix} 1 & 2 & -1 \\ 4 & 3 & 1 \\ 2 & 2 & 3 \end{pmatrix}\begin{pmatrix} x_1 \\ x_2 \\ x_3 \end{pmatrix} = \begin{pmatrix} 2 \\ 3 \\ 5 \end{pmatrix}$$

The following steps show how to apply the direct method to solve this problem.

Listing 8.10

```
>>A = [1, 2, -1; 4, 3, 1; 2, 2,3];
>> [L, U] = ludecomp(A);
>>f = [2; 3; 5];
>> ludirect(L,U, f)
ans =
-1.0000
2.0000
1.0000
```

8.6.2 Error estimation for linear systems

Estimation of the error vector **e**=**x**−**x*** is a delicate task, since we usually have no way of computing this vector. It is possible, at least within the limits of machine precision, to compute the residual vector **r**=**Ax**−**f**, which vanishes if **e**=**o**. However, a small residual does not guarantee that the error is small. Let us analyze the relationship between **e** and **r** more generally. Observe that $\mathbf{r}=\mathbf{Ax}-\mathbf{Ax}^*=\mathbf{A}(\mathbf{x}-\mathbf{x}^*)=\mathbf{Ae}$, that is, $\mathbf{e}=\mathbf{A}^{-1}\mathbf{r}$. Then the ratio $||\mathbf{e}||/||\mathbf{f}||$ is

$$\frac{\|\mathbf{e}\|}{\|\mathbf{f}\|}=\frac{\|\mathbf{e}\|}{\|\mathbf{Ax}^*\|}=\frac{\|\mathbf{A}^{-1}\mathbf{r}\|}{\|\mathbf{Ax}^*\|}\le\frac{\|\mathbf{A}^{-1}\|\cdot\|\mathbf{r}\|}{\|\mathbf{Ax}^*\|}.$$

But $||\mathbf{Ax}^*||\le||\mathbf{A}||\cdot||\mathbf{x}^*||$, so we can replace the denominator on the left side and multiply through by $||\mathbf{A}||$ to obtain the estimate

$$\frac{\|\mathbf{e}\|}{\|\mathbf{x}^*\|}\le\|\mathbf{A}\|\cdot\|\mathbf{A}^{-1}\|\frac{\|\mathbf{r}\|}{\|\mathbf{f}\|}=\mathrm{cond}(\mathbf{A})\frac{\|\mathbf{r}\|}{\|\mathbf{f}\|}.$$

This inequality asserts that the relative error is small when the relative magnitude of the residual is small and cond(**A**) is not too large.

8.7 Iterative Methods

Iterative methods for $\mathbf{Ax}=\mathbf{f}$ are infinite methods and they produce only approximate results. They are easy to define and can be highly efficient. However, a thorough study of the problem is needed in order to choose an appropriate method.

8.7.1 Basic principles of iterative methods

Iterative methods are usually based on the following equivalent form of system (8.1):

$$\mathbf{x}=\mathbf{Bx}+\mathbf{g} \tag{8.10}$$

where **B** is an iterative matrix. We construct this form by analogy with (7.1), simply adapting the same principle to system (8.1). The components of the iterative matrix are expressed in terms of the components of matrix **A**. Vector **g** is also expressed in terms of matrix **A** and vector **f**. It is often not practical to calculate **B** explicitly but it is implied that in principle this can be done. On the basis of form (8.10), an iterative process is constructed as follows:

1) Choose an initial vector $\mathbf{x}^{(0)}$.
2) Compute successively

$$\mathbf{x}^{(k+1)}=\mathbf{Bx}^{(k)}+\mathbf{g} \tag{8.11}$$

or

$$\mathbf{x}^{(k+1)}=\mathbf{B}_k\mathbf{x}^{(k)}+\mathbf{g}_k,\ k=0,\ 1,\ \dots, \tag{8.12}$$

3) Conclude the iterative process if

$$\frac{\left\|\mathbf{x}^{(k+1)}-\mathbf{x}^{(k)}\right\|}{\left\|\mathbf{x}^{(k+1)}\right\|}\le\varepsilon_p$$

or

$$\frac{\left\|\mathbf{r}^{(k+1)}\right\|}{\left\|\mathbf{x}^{(k+1)}\right\|}=\frac{\left\|\mathbf{Ax}^{(k+1)}-\mathbf{f}\right\|}{\left\|\mathbf{x}^{(k+1)}\right\|}\le\varepsilon_p.$$

Then

$$\frac{\left\|\mathbf{x}^{(k+1)}-\mathbf{x}^*\right\|}{\left\|\mathbf{x}^*\right\|} \leq \varepsilon_p .$$

Iterative process (8.11) is called stationary and process (8.12) is accordingly called nonstationary. How are iterations $\mathbf{x}^{(k)}$ related to the exact answer for problem (8.1)? To clear up this question, let us carry out a simple analysis. Every approximation $\mathbf{x}^{(k)}$ may be represented in the following form:

$$\begin{aligned}\mathbf{x}^{(k)} &= \mathbf{x}^* + \mathbf{e}^{(k)}, \\ \mathbf{x}^{(k+1)} &= \mathbf{x}^* + \mathbf{e}^{(k+1)},\end{aligned} \tag{8.13}$$

where $\mathbf{e}^{(k)}$ is the error vector of kth approximation. After substituting (8.13) into (8.11) and taking into account (8.10), we obtain

$$\mathbf{e}^{(k+1)} = \mathbf{B}\mathbf{e}^{(k)} \text{ or } \mathbf{e}^{(k)} = \mathbf{B}^k\mathbf{e}^{(0)}, \; k=0,\, 1,\, \ldots . \tag{8.14}$$

If

$$\lim_{k\to\infty}\left\|\mathbf{e}^{(k)}\right\| = 0,$$

then the iterations converge and

$$\lim_{k\to\infty}\mathbf{x}^{(k)} = \mathbf{x}^* .$$

Theorem 8.2

A necessary and sufficient condition for iterative process (8.11) to be convergent for an arbitrary initial vector $\mathbf{x}^{(0)}$ is

$$s(\mathbf{B}) < 1, \tag{8.15}$$

here $s(\mathbf{B})=\max|\lambda_n(\mathbf{B})|$ is the spectral radius of the matrix $\mathbf{B}$.

There is another sufficient condition that provides convergence: if $||\mathbf{B}||<1$, then condition (8.15) is satisfied, because $s(\mathbf{B})\leq||\mathbf{B}||$ for any matrix norm.

Iterative process (8.13) is connected with the algorithm and rounding errors; the perturbation error does not occur. The algorithm error is easily estimated. For simplicity, the iteration is started with $\mathbf{x}^{(0)}=\mathbf{g}$, then

$$\mathbf{x}^{(k)} = \sum_{l=0}^{k}\mathbf{B}^l\mathbf{g}, \; \mathbf{x}^* = \sum_{l=0}^{\infty}\mathbf{B}^l\mathbf{g}.$$

Hence

$$\left\|\mathbf{e}_a^{(k)}\right\| = \left\|\mathbf{x}^* - \mathbf{x}^{(k)}\right\| = \left\|\sum_{l=k+1}^{k}\mathbf{B}^l\mathbf{g}\right\| \leq \sum_{l=k+1}^{k}\|\mathbf{B}\|^l\,\|\mathbf{g}\| .$$

For $||\mathbf{B}||<1$ we obtain

$$\left\|\mathbf{e}_a^{(k)}\right\| \leq \frac{\|\mathbf{B}\|^{k+1}\,\|\mathbf{g}\|}{1-\|\mathbf{B}\|},$$

so the smaller the spectral radius of the matrix $\mathbf{B}$, the faster approximations $\mathbf{x}^{(k)}$ approach $\mathbf{x}^*$. As for the rounding errors, the following estimate may be obtained:

$$\left\| \mathbf{e}_r^{(k)} \right\| \le \frac{\varepsilon_1 \|\mathbf{g}\|}{1 - \|\mathbf{B}\|}, \text{ for all } k.$$

As we can see, iterative methods are very stable in terms of rounding errors.

8.7.2 Jacobi method

Let us represent a matrix **A** as

$$\mathbf{A} = \mathbf{L} + \mathbf{D} + \mathbf{U}, \tag{8.16}$$

where

$$\mathbf{L} = \begin{pmatrix} 0 & & & & \\ * & \cdot & & 0 & \\ * & * & \cdot & & \\ * & \cdot & * & \cdot & \\ * & * & * & * & 0 \end{pmatrix}, \ \mathbf{U} = \begin{pmatrix} 0 & * & * & * & * \\ & \cdot & * & \cdot & * \\ & & \cdot & * & * \\ & 0 & & \cdot & * \\ & & & & 0 \end{pmatrix},$$

$$\mathbf{D} = \operatorname{diag}(a_{11}, \ldots, a_{NN}).$$

Now we can construct an equivalent form (8.10) in the following manner

$$\mathbf{Ax} = (\mathbf{L} + \mathbf{D} + \mathbf{U})\mathbf{x} = \mathbf{f},$$

$$\mathbf{Dx} = -(\mathbf{L} + \mathbf{U})\mathbf{x} + \mathbf{f} = (\mathbf{D} - \mathbf{A})\mathbf{x} + \mathbf{f},$$

and finally

$$\mathbf{x} = (\mathbf{I} - \mathbf{D}^{-1}\mathbf{A})\mathbf{x} + \mathbf{D}^{-1}\mathbf{f}. \tag{8.17}$$

Therefore, iterative matrix **B** and vector **g** for the Jacobi method is

$$\mathbf{B} = \mathbf{B}_J = \mathbf{I} - \mathbf{D}^{-1}\mathbf{A}, \ \mathbf{g} = \mathbf{g}_J = \mathbf{D}^{-1}\mathbf{f}.$$

The construction of equivalent form (8.17) consists of solving the *n*th equation explicitly for the *n*th unknown and putting everything else on the right-hand side. Because matrix $\mathbf{B}_J$ is easily calculated, one can estimate $s(\mathbf{B}_J)$ to determine if the Jacobi method is convergent. There are also sufficient conditions which allow for clarifying convergence of the Jacobi method immediately in terms of matrix **A**. They are: (1) **A** is a matrix with strong diagonal dominance, so taking into account the structure of $\mathbf{B}_J$, it is easy to show that $||\mathbf{B}_J||_\infty<1$; and (2) **A** is an irreducible matrix with weak diagonal dominance.

The implementation of the Jacobi method requires two utility functions, *ldu* and *invdiag*. The function *ldu* decomposes a given matrix into lower triangular, diagonal and upper triangular matrices.

Listing 8.11

```
function [L, D, U] = ldu(A)
% ldu function decompose a matrix into L, D & U matrices
% such that A = L + D + U
% input:  A - input square matrix
% output: L - lower triangular factor
```

```
%          D - diagonal factor
%          U - upper triangular factor
n = length(A); L = zeros(n); D = zeros(n); U = zeros(n);
for i=1:n
    for j=1:n
        if i<j
            U(i, j) = A(i, j);
        elseif i > j
            L(i, j) = A(i, j);
        else
            D(i, j) = A(i, j);
        end;
    end;
end;
```

The *invdiag* function returns the inverse of a given diagonal matrix. It was important to implement as a separate and efficient function rather than using the built-in *inv* function since the diagonal matrix is sparsely populated and it should not take that much computational effort.

Listing 8.12

```
function Dn = invdiag(D)
% invdiag function comptes the inverse of a diagonal matrix
% input:  D - diagonal matrix
% output: Dn - diagonal inverse matrix
n = length(D); Dn = zeros(n);
for i=1:n
    Dn(i, i) = 1/D(i,i);
end;
```

The following code listing implements the Jacobi method using the above utility functions. It takes three input arguments; the coefficient matrix, right-hand side vector and error tolerance.

Listing 8.13

```
function x = jacobi(A, f, tol)
% jacobi function solves a linear system of equations using Jacobi method
% input:  A - input coefficient matrix
%         f - right-hand side vector
% output: x - solution of linear system of equations
n = length(A);
x0 = zeros(n,1);
I = eye(n);
[L, D, U] = ldu(A);
invD = invdiag(D);
```

```
x = (I-invD*A)*x0 + invD*f;
i = 0;
while norm(x-x0) > tol
    i = i + 1;
    x0 = x;
    x = (I-invD*A)*x0 + invD*f;
end;
s = sprintf('Total number of iterations = %d', i);
disp(s)
```

We have the following system of equations given in a matrix form.

$$\begin{pmatrix} 2 & -1 & 0 \\ -1 & 2 & -1 \\ 0 & -1 & 2 \end{pmatrix} \begin{pmatrix} x_1 \\ x_2 \\ x_3 \end{pmatrix} = \begin{pmatrix} 1/3 \\ 1 \\ -1/3 \end{pmatrix}$$

The following steps show how to solve it using the Jacobi method.

Listing 8.14

```
>> A = [2 -1 0; -1 2 -1; 0 -1 2];
>> f = [1/3; 1; -1/3];
>> x = jacobi(A, f, 1e-6)
Total number of iterations = 38
x =
   0.6667
   1.0000
   0.3333
```

It takes 38 iterations to reach the solution with desired accuracy.

8.7.3 Gauss-Seidel method

Let us refer again to representation (8.18). We can construct an equivalent form (8.12) in a different manner:

$$\mathbf{Ax} = (\mathbf{L} + \mathbf{D} + \mathbf{U})\mathbf{x} = \mathbf{f}\,,$$

$$(\mathbf{L} + \mathbf{D})\mathbf{x} = -\mathbf{Ux} + \mathbf{f}\,,$$

$$(\mathbf{B} = \mathbf{B}_{GS} = -(\mathbf{L} + \mathbf{D})^{-1}\mathbf{U}\,,\ \mathbf{g} = \mathbf{g}_{GS} = -(\mathbf{L} + \mathbf{D})^{-1}\mathbf{f}\,)$$

or

$$(\mathbf{U} + \mathbf{D})\mathbf{x} = -\mathbf{Lx} + \mathbf{f}\,,$$

$$(\mathbf{B} = \mathbf{B}_{GS} = -(\mathbf{U} + \mathbf{D})^{-1}\mathbf{L}\,, \mathbf{g} = \mathbf{g}_{GS} = -(\mathbf{U} + \mathbf{D})^{-1}\mathbf{f}\,).$$

We do not need to invert the matrices **L+D** or **U+D**, because we can immediately apply the iteration

$$(\mathbf{L} + \mathbf{D})\mathbf{x}^{(k+1)} = -\mathbf{Ux}^{(k)} + \mathbf{f} \tag{8.18}$$

or

$$(\mathbf{U}+\mathbf{D})\mathbf{x}^{(k+1)} = -\mathbf{L}\mathbf{x}^{(k)} + \mathbf{f}\,. \tag{8.19}$$

The matrix **L**+**D** is a lower triangular one and **U**+**D** is an upper triangular matrix. Therefore, systems (8.18) and (8.19) are easily solved with respect to $\mathbf{x}^{(k+1)}$. Because the explicit calculation of matrix $\mathbf{B}_{GS}$ is impractical, the sufficient convergence conditions are of importance. They are: (1) **A** is a matrix with strong diagonal dominance; (2) **A** is an irreducible matrix with weak diagonal dominance; and (3) **A** is a SPD matrix. In this last case, if the Jacobi method converges ($s(\mathbf{B}_J)<1$), then the Gauss-Seidel method has more rapid convergence because $s(\mathbf{B}_{GS})=s^2(\mathbf{B}_J)$ for this type of matrix.

The following function implements the Gauss-Siedel's iterative method to solve a system of linear equations. The function signature is same as the Jacobi method.

Listing 8.15

```
function x = gauss(A, f, tol)
% gauss function solves a linear system of equations using Gauss Seidel method
% input:   A - input coefficient matrix
%          f - right-hand side vector
% output:  x - solution of linear system of eqs
n = length(A);
x0 = zeros(n,1);
[L, D, U] = ldu(A);
f1 = -L * x0 + f;
x = bsubstt(U+D, f1);
i = 0;
while norm(x-x0) > tol
     i = i + 1;
     x0 = x;
     f1 = -L * x0 + f;
     x = bsubstt(U+D, f1);
end;
s = sprintf('Total number of iterations = %d', i);
disp(s);
```

In the following, we apply the Gauss-Siedel method to the problem in last example.

Listing 8.16

```
>> x = gauss(A, f, 1e-6)
Total number of iterations = 20
x =
   0.6667
   1.0000
   0.3333
```

It shows that the Gauss-Siedel method is superior to the Jacobi method since it takes only 20 iterations to solve the same problem.

8.7.4 Method of relaxation

When employing an iterative method, the main question is how to achieve rapid convergence, or to put this in other words, how to construct an iterative matrix with the smallest $s(\mathbf{B})$ possible. To do this, one should be able to exert control over the properties of iterative matrix **B**. This can be realized with the use of a special procedure called the method of relaxation.

To begin with, let us discuss the general scheme of this method. Consider system (8.1) with an SPD matrix **A**. For an arbitrary vector $\mathbf{x}^{(k)}$, which is considered as an approximate solution, we define an error function

$$\varphi(\mathbf{x}^{(k)}) = \left(\mathbf{x}^{(k)} - \mathbf{x}^*, \mathbf{A}(\mathbf{x}^{(k)} - \mathbf{x}^*)\right) = (\mathbf{e}^{(k)}, \mathbf{A}\mathbf{e}^{(k)}) = (\mathbf{e}^{(k)}, \mathbf{r}^{(k)}).$$

In order to define another vector $\mathbf{x}^{(k+1)}$ which differs from $\mathbf{x}^{(k)}$ only in the nth component such that $\varphi(\mathbf{x}^{(k+1)})<\varphi(\mathbf{x}^{(k)})$, we put $\mathbf{x}^{(k+1)}=\mathbf{x}^{(k)}+\alpha\mathbf{b}_n$ ($\mathbf{b}_n$ is the basis vector, see 8.1). Then, it is easy to obtain

$$\varphi(\mathbf{x}^{(k+1)}) = \varphi(\mathbf{x}^{(k)}) + \frac{(\alpha a_{nn} + r_n^{(k)})^2}{a_{nn}} - \frac{(r_n^{(k)})^2}{a_{nn}}, \tag{8.20}$$

where $(\mathbf{r}^{(k)})_n$ is the nth component of the residual vector: $(\mathbf{r}^{(k)})_n=(\mathbf{A}\mathbf{x}^{(k)}-\mathbf{f}, \mathbf{b}_n)$. The sum of the second and third terms in (8.20) can be made negative; we simply choose α so that

$$\left|\alpha a_{nn} + r_n^{(k)}\right| < \left|r_n^{(k)}\right|$$

or

$$\alpha = -\omega\frac{r_n^{(k)}}{a_{nn}}, 0<\omega<2,$$

and

$$x_n^{(k+1)} = x_n^{(k)} - \omega\frac{r_n^{(k)}}{a_{nn}},$$

where ω is called the relaxation parameter. By choosing parameter α for $n=1, \ldots, N$ we arrive at the following iterative scheme:

$$\begin{aligned}&\mathbf{x}^{(k+1)} = \mathbf{x}^{(k)} - \omega\mathbf{D}^{-1}(\mathbf{A}\mathbf{x}^{(k)} - \mathbf{f}) = \\ &((1-\omega)\mathbf{I} + \omega\mathbf{B}_\mathrm{J})\mathbf{x}^{(k)} + \omega\mathbf{D}^{-1}\mathbf{f} = \\ &((1-\omega)\mathbf{I} + \omega\mathbf{B}_\mathrm{J})\mathbf{x}^{(k)} + \omega\mathbf{D}^{-1}\mathbf{f} = \mathbf{B}_\mathrm{JR}(\omega)\mathbf{x}^{(k)} + \mathbf{g}_\mathrm{JR}(\omega)\end{aligned} \tag{8.21}$$

The equivalent form (8.10) for this iterative process may be also written as

$$\mathbf{D}\mathbf{x} = \mathbf{D}\mathbf{x} - \omega(\mathbf{A}\mathbf{x} - \mathbf{f}). \tag{8.22}$$

Now we have an iterative matrix that depends on a parameter ω. One can find such ω that provides the minimal value of $s(\mathbf{B}_\mathrm{JR}(\omega))$. This parameter is called the optimal relaxation parameter and is denoted as ω_opt. The method of relaxation (8.21) is based on the Jacobi method ($\mathbf{B}_\mathrm{JR}(1)=\mathbf{B}_\mathrm{J}$). It is easy to see that $\lambda(\mathbf{B}_\mathrm{JR}(\omega))=1+\omega(\lambda(\mathbf{B}_\mathrm{J})-1)$. Let $-\lambda_l\le\lambda(\mathbf{B}_\mathrm{J})\le\lambda_r$ then the optimal relaxation parameter is

$$\omega_{\text{opt}} = \frac{2}{2 + \lambda_l - \lambda_r}$$

and

$$s(\mathbf{B}_{\text{JR}}(\omega_{\text{opt}})) = \frac{\lambda_l + \lambda_r}{2 + \lambda_l - \lambda_r}\,.$$

The next logical step is to construct the relaxation method based on the Gauss-Seidel iteration (8.18). Taking into account representation (8.16), one can transform (8.22) into the following equivalent system:

$$\mathbf{Dx} = \mathbf{Dx} - \omega(\mathbf{L} + \mathbf{D} + \mathbf{U})\mathbf{x} + \omega\mathbf{f}\,,$$
$$(\mathbf{D} + \omega\mathbf{L})\mathbf{x} = (\mathbf{D} - \omega(\mathbf{D} + \mathbf{U}))\mathbf{x} + \omega\mathbf{f}\,,$$
$$(\,\mathbf{B}_{\text{SOR}} = (\mathbf{D} + \omega\mathbf{L})^{-1}(\mathbf{D} - \omega(\mathbf{D} + \mathbf{U}))\,).$$

As before, we do not need to invert matrix $\mathbf{D}+\omega\mathbf{L}$ and the iterative scheme is

$$(\mathbf{D} + \omega\mathbf{L})\mathbf{x}^{(k+1)} = (\mathbf{D} - \omega(\mathbf{D} + \mathbf{U}))\mathbf{x}^{(k)} + \omega\mathbf{f}\,, \qquad (8.23)$$
$$k=0,\ 1,\ \ldots.$$

Matrix $\mathbf{D}+\omega\mathbf{L}$ is a lower triangular one, therefore system (8.23) can be easily solved with respect to $\mathbf{x}^{(k+1)}$. It was noticed in practice (in some cases it was proved) that convergence of iterative scheme (8.23) with $\omega<1$ (under-relaxation) is slower in comparison to $\omega\geq1$ (over-relaxation). Thus, $\omega\geq1$ is usually used and iterative scheme (8.23) is called the method of successive over-relaxation (SOR). There is only one procedure, proposed by David Young, to calculate the optimal relaxation parameter ω. Let us briefly consider this result.

The Jacobi matrix $\mathbf{B}_J=\mathbf{I}-\mathbf{D}^{-1}\mathbf{A}=-\mathbf{D}^{-1}\mathbf{L}-\mathbf{D}^{-1}\mathbf{U}$ is consistently ordered if

$$\lambda(-\alpha\mathbf{D}^{-1}\mathbf{L} - \frac{1}{\alpha}\mathbf{D}^{-1}\mathbf{U}) = \lambda(\mathbf{B}_J)$$

for any $\alpha\neq0$. For example, block tridiagonal matrices of the form

$$\mathbf{T} = \begin{pmatrix} \mathbf{D}_1 & \mathbf{U}_1 & & & & \\ \mathbf{L}_2 & \mathbf{D}_2 & \mathbf{U}_2 & & 0 & \\ & \cdot & \cdot & \cdot & & \\ & & \cdot & \cdot & \cdot & \\ & 0 & & \mathbf{L}_{p-1} & \mathbf{D}_{p-1} & \mathbf{U}_{p-1} \\ & & & & \mathbf{L}_p & \mathbf{D}_p \end{pmatrix},$$

where $\mathbf{D}_m$ is a square matrix of size $N_m\times N_m$ ($m=1,\ \ldots,\ p$), $\mathbf{L}_m$ and $\mathbf{U}_{m-1}$ are matrices of size $N_m\times N_{m+1}$, and $N_{m+1}\times N_m$ respectively ($m=2,\ \ldots,\ p;\ N_1+\ldots+N_p=N$) are consistently ordered. Let $\mathbf{A}$ be an SPD consistently ordered matrix. If the Jacobi method converges ($s(\mathbf{B}_J)<1$), then the optimal relaxation parameter for the SOR is calculated by the formula:

$$\omega_{\mathrm{opt}} = \frac{2}{1+\sqrt{1-s^2(\mathbf{B}_\mathrm{J})}} \text{ and } s(\mathbf{B}_{\mathrm{SOR}}) = \omega_{\mathrm{opt}} - 1. \tag{8.24}$$

Calculating the spectral radius of the Jacobi matrix $\mathbf{B}_\mathrm{J}$ requires an impractical amount of computation, so this is seldom done. However, relatively inexpensive rough estimates of $s(\mathbf{B}_\mathrm{J})$ can yield reasonable estimates for the optimal value of ω_{opt}.

As for other types of matrices, we do not know how to determine the optimal relaxation parameter. However, there are some conditions which are sufficient for the convergence of the SOR. First, if **A** is an SPD matrix, then the SOR converges when $0<\omega<2$. In this case the relaxation parameter given by (8.24) is still close to optimal, because $\omega_{\mathrm{opt}}-1 \leq s(\mathbf{B}_{\mathrm{SOR}}) \leq (\omega_{\mathrm{opt}}-1)^{1/2}$. Second, if $\boldsymbol{A}$ is a matrix with strong diagonal dominance, or with weak diagonal dominance and irreducible, then some $\omega^*>1$ exists. The SOR converges when $0<\omega<\omega^*$.

Successive Over Relaxation or SOR method introduces a parameter to control the convergence of the iterative process, as shown in the following implementation. The signature of following function is same as previous functions except the addition of *w* control parameter.

Listing 8.17
```
function x = sor(A, f, w, tol)
% gauss function solves a linear system of equations using Gauss Seidel method
% input:   A - input coefficient matrix
%          f - right-hand side vector
%          w - omega relaxation parameter
%          tol - error tolerance
% output: x - solution of linear system of equations
n = length(A);
x0 = zeros(n,1);
[L, D, U] = ldu(A);
f1 = (D-w*(D + U))*x0 + w*f;
x = fsubstt(D + w*L, f1);
i = 0;
while norm(x-x0) > tol
    i = i + 1;
    x0 = x;
    f1 = (D-w*(D + U))*x0 + w*f;
    x = fsubstt(D + w*L, f1);
end;
s = sprintf('Total number of iterations = %d', i);
disp(s);
```

We apply the SOR method to the same problem discussed in previous sections. There are a number of methods to find the optimal value of control parameter, *w*. For the sake of simplicity, we have arbitrarily chosen the value in the following.

Readers are encouraged to try different values to see how it impacts the convergence speed.

Listing 8.18

```
>> x = sor(A, f, 1.2, 1e-6)
Total number of iterations = 9
x =
   0.6667
   1.0000
   0.3333
```

It takes only 9 iterations to solve the same problem which shows that SOR is the best iterative method to solve a system of linear equations.

8.7.5 Variational iterative methods

It is evident from the foregoing account that, in general, we need some information about the spectral radius of an iterative matrix in order to analyze convergence of an iterative method. Often, it is impractical to calculate or estimate $s(\mathbf{B})$. There are methods wherein convergence is ensured by special parameters, which in turn depend on an approximate solution. It is clear that the realization of such methods gives rise to non-stationary iterative processes (8.12).

We shall consider an iterative scheme of the following form:

$$\mathbf{x}^{(k+1)} = \mathbf{x}^{(k)} - \tau_k(\mathbf{A}\mathbf{x}^{(k)} - \mathbf{f}),\ k{=}0,\ 1,\ \dots, \tag{8.25}$$

where $\tau_k \neq 0$ is an iterative parameter. Let us introduce residual $\mathbf{r}^{(k)}=\mathbf{A}\mathbf{x}^{(k)}-\mathbf{f}$. Using (8.25), the residual vector $\mathbf{r}^{(k+1)}$ associated with the next approximation can be expressed in terms of $\mathbf{r}^{(k)}$

$$\mathbf{A}\mathbf{x}^{(k+1)} - \mathbf{f} = (\mathbf{A}\mathbf{x}^{(k)} - \mathbf{f}) - \tau_k\mathbf{A}(\mathbf{A}\mathbf{x}^{(k)} - \mathbf{f}) =$$

$$\mathbf{r}^{(k)} - \tau_k\mathbf{A}\mathbf{r}^{(k)} = \mathbf{r}^{(k+1)}$$

Thus, we can determine parameter $\tau_k \neq 0$ in such a manner that $||\mathbf{r}^{(k+1)}||_2$ will be minimal. The norm of residual $\mathbf{r}^{(k+1)}$ is expressed as follows:

$$\left\|\mathbf{r}^{(k+1)}\right\|_2^2 = (\mathbf{r}^{(k+1)}, \mathbf{r}^{(k+1)}) =$$

$$(\mathbf{r}^{(k)} - \tau_k\mathbf{A}\mathbf{r}^{(k)}, \mathbf{r}^{(k)} - \tau_k\mathbf{A}\mathbf{r}^{(k)}) = \tag{8.26}$$

$$(\mathbf{r}^{(k)}, \mathbf{r}^{(k)}) - 2\tau_k(\mathbf{r}^{(k)}, \mathbf{A}\mathbf{r}^{(k)}) + \tau_k^2(\mathbf{A}\mathbf{r}^{(k)}, \mathbf{A}\mathbf{r}^{(k)}) = \varphi(\tau_k)$$

Condition $\varphi'(\tau_k)=0$ is the condition of the minimum of $\varphi(\tau_k)$, because

$$\varphi''(\tau_k) = (\mathbf{A}\mathbf{r}^{(k)}, \mathbf{A}\mathbf{r}^{(k)}) = \left\|\mathbf{A}\mathbf{r}^{(k)}\right\|_2^2 > 0\,.$$

Then τ_k can be calculated as

$$\tau_k = \frac{(\mathbf{r}^{(k)}, \mathbf{A}\mathbf{r}^{(k)})}{(\mathbf{A}\mathbf{r}^{(k)}, \mathbf{A}\mathbf{r}^{(k)})}\,. \tag{8.27}$$

Iterative schemes (8.25) and (8.27) together comprise the method of minimal residual. Let us analyze convergence of this method. If we substitute (8.27) into (8.26), then

$$\left\|\mathbf{r}^{(k+1)}\right\|_2^2 = \left\|\mathbf{r}^{(k)}\right\|_2^2 - \frac{(\mathbf{r}^{(k)}, \mathbf{A}\mathbf{r}^{(k)})^2}{(\mathbf{A}\mathbf{r}^{(k)}, \mathbf{A}\mathbf{r}^{(k)})} > 0. \tag{8.28}$$

Both terms in the right-hand side of (8.28) are positive, therefore

$$\left\|\mathbf{r}^{(k+1)}\right\|_2^2 < \left\|\mathbf{r}^{(k)}\right\|_2^2$$

for any definite matrix **A** and this proves convergence of iterative scheme (8.25). Now consider the rate of convergence of the method of minimal residual. Suppose a matrix **A** of system (8.1) has the following properties:

$$\gamma_1 \le \frac{(\mathbf{x}, \mathbf{A}\mathbf{x})}{(\mathbf{x}, \mathbf{x})} \le \gamma_2, \mathbf{x}\in R^N, 0<\gamma_1\le\gamma_2<\infty,$$

and

$$\left\|\mathbf{A} - \mathbf{A}^T\right\| \le 2\gamma_3.$$

Then

$$\left\|\mathbf{r}^{(k+1)}\right\|_2 = q\left\|\mathbf{r}^{(k)}\right\|_2$$

with

$$q = \frac{\gamma_2^2 - \gamma_1^2 + 4\gamma_3\sqrt{\gamma_3^2 + \gamma_1\gamma_2}}{(\gamma_1 + \gamma_2)^2 + 4\gamma_3^2} < 1.$$

If all eigenvalues of matrix **A** are real, then we can put

$$\gamma_2 = \max_n \lambda_n(\mathbf{A})$$

and

$$\gamma_1 = \min_n \lambda_n(\mathbf{A}).$$

For SPD matrices one can use iterative scheme (8.25) and calculate τ_k by other means. Instead of the residual, let us consider the actual error $\mathbf{e}^{(k)}=\mathbf{x}^{(k)}-\mathbf{x}^*$. Taking into account that $\mathbf{A}\mathbf{x}^*=\mathbf{f}$, from (8.25) we obtain

$$\begin{aligned}\mathbf{e}^{(k+1)} &= \mathbf{x}^{(k+1)} - \mathbf{x}^* = \mathbf{x}^{(k)} - \mathbf{x}^* + \tau_k\mathbf{A}\left(\mathbf{x}^{(k)} - \mathbf{x}^*\right) = \\ &\mathbf{e}^{(k)} - \tau_k\mathbf{A}\mathbf{e}^{(k)}\end{aligned} \tag{8.29}$$

We have no way of calculating the actual error, but

$$\mathbf{A}\mathbf{e}^{(k+1)} = \mathbf{A}\mathbf{x}^{(k)} - \mathbf{A}\mathbf{x}^* = \mathbf{A}\mathbf{x}^{(k)} - \mathbf{f} = \mathbf{r}^{(k)}.$$

So we can determine τ_k by minimizing the value of $(\mathbf{e}^{(k+1)}, \mathbf{A}\mathbf{e}^{(k+1)})$. Using (8.29) we obtain

$$(\mathbf{e}^{(k+1)}, \mathbf{A}\mathbf{e}^{(k+1)}) = (\mathbf{e}^{(k)} - \tau_k\mathbf{r}^{(k)}, \mathbf{r}^{(k)} - \tau_k\mathbf{A}\mathbf{r}^{(k)}) =$$

$$(\mathbf{e}^{(k)}, \mathbf{r}^{(k)}) - 2\tau_k(\mathbf{r}^{(k)}, \mathbf{r}^{(k)}) + \tau_k^2(\mathbf{r}^{(k)}, \mathbf{A}\mathbf{r}^{(k)}) = \varphi(\tau_k)$$

Condition $\varphi'(\tau_k)=0$ is the condition of the minimum of $\varphi(\tau_k)$, because $\varphi''(\tau_k)=(\mathbf{r}^{(k)}, \mathbf{A}\mathbf{r}^{(k)})>0$ (**A** is an SPD matrix). So τ_k is determined as follows:

$$\tau_k = \frac{(\mathbf{r}^{(k)}, \mathbf{r}^{(k)})}{(\mathbf{r}^{(k)}, \mathbf{A}\mathbf{r}^{(k)})}. \tag{8.30}$$

Iterative scheme (8.25) with τ_k from (8.30) is called the method of steepest descent. Properties of this method are similar to the method of minimal residual.

On the basis of the method of steepest descent we can construct an iterative method with more rapid convergence. Let us modify (8.25) and write it in the following form:

$$\mathbf{x}^{(k+1)} = \mathbf{x}^{(k)} - \tau_k \mathbf{d}^{(k)}, \; k=0, 1, \ldots . \tag{8.31}$$

Here $\mathbf{d}^{(k)}$ is a direction vector which points towards the solution of a linear system. In the method of steepest descent, direction vectors are residuals $\mathbf{r}^{(k)}$, but this choice of directions turns out to be a poor one for many systems. It is possible to choose better directions by computing direction vectors as

$$\mathbf{d}^{(0)} = \mathbf{r}^{(0)}, \; \mathbf{d}^{(k)} = \mathbf{r}^{(k)} + \alpha_k \mathbf{d}^{(k-1)}, \; k=1, 2, \ldots . \tag{8.32}$$

Parameters α_k are determined in such a way as to provide orthogonality of vectors $\mathbf{d}^{(k)}$ and $\mathbf{A}\mathbf{d}^{(k-1)}$ $((\mathbf{d}^{(k)}, \mathbf{A}\mathbf{d}^{(k-1)})=0)$, then

$$\alpha_k = -\frac{(\mathbf{r}^{(k)}, \mathbf{A}\mathbf{d}^{(k-1)})}{(\mathbf{d}^{(k-1)}, \mathbf{A}\mathbf{d}^{(k-1)})}. \tag{8.33}$$

Using the same considerations as those for the method of steepest descent, τ_k is determined so that it provides a minimal value of $(\mathbf{e}^{(k+1)}, \mathbf{A}\mathbf{e}^{(k+1)})$. Then we have

$$\tau_k = \frac{(\mathbf{r}^{(k)}, \mathbf{d}^{(k)})}{(\mathbf{d}^{(k)}, \mathbf{A}\mathbf{d}^{(k)})}. \tag{8.34}$$

Iterative scheme (8.31) with equations (8.32), (8.33), and (8.34) comprise the conjugate gradient method. The error of the kth approximation generated by this method obeys the following bound

$$(\mathbf{e}^{(k)}, \mathbf{A}\mathbf{e}^{(k)}) \le 2(\mathbf{e}^{(0)}, \mathbf{A}\mathbf{e}^{(0)})(q(\mathbf{A}))^k$$

where

$$q(\mathbf{A}) = \frac{\sqrt{\max\limits_n \lambda_n(\mathbf{A})} - \sqrt{\min\limits_n \lambda_n(\mathbf{A})}}{\sqrt{\max\limits_n \lambda_n(\mathbf{A})} + \sqrt{\min\limits_n \lambda_n(\mathbf{A})}} < 1.$$

The distinctive feature of this method is that, ignoring round-off errors, we obtain $\mathbf{r}^{(k)}=0$ at some $k \le N$, therefore $\mathbf{x}^{(k)}=\mathbf{x}^*$. By this means we need no more than N iterations to achieve the true solution. For large systems of equations this is not such a promising result. Nevertheless, in some cases the conjugate gradient method and its modifications produce very rapid convergence.

8.8 Exercises

The problems consist of two parts:

1) solve a system of linear equations using LU-factorization (systems 1–20),
2) solve a system of linear equations using Gauss-Seidel method (systems 21–30).

Estimate the error of approximate solution. Methodical comments: to compute the LU-factorization use the function `[L,U]=ludecomp(A);` to estimate the condition number use the function `cond(A)`.
Set of systems of equations to solve:

1. $$\begin{cases} 0.63x_1 + 1.00x_2 + 0.71x_3 + 0.34x_4 = 2.08 \\ 1.17x_1 + 0.18x_2 - 0.65x_3 + 0.71x_4 = 0.17 \\ 2.71x_1 - 0.75x_2 + 1.17x_3 - 2.35x_4 = 1.28 \\ 3.58x_1 + 0.21x_2 - 3.45x_3 - 1.18x_4 = 0.05 \end{cases}$$

2. $$\begin{cases} 3.51x_1 + 0.17x_2 + 3.75x_3 - 0.28x_4 = 0.75 \\ 4.52x_1 + 2.11x_2 - 0.11x_3 - 0.12x_4 = 1.11 \\ -2.11x_1 + 3.17x_2 + 0.12x_3 - 0.15x_4 = 0.21 \\ 3.17x_1 + 1.81x_2 - 3.17x_3 + 0.22x_4 = 0.05 \end{cases}$$

3. $$\begin{cases} 0.17x_1 + 0.75x_2 - 0.18x_3 + 0.21x_4 = 0.11 \\ 0.75x_1 + 0.13x_2 + 0.11x_3 + 1.00x_4 = 2.00 \\ -0.33x_1 + 0.11x_2 + 3.01x_3 - 2.01x_4 = 0.11 \\ 0.11x_1 + 1.12x_2 + 1.11x_3 - 1.31x_4 = 0.13 \end{cases}$$

4. $$\begin{cases} -1.00x_1 + 0.13x_2 - 2.00x_3 - 0.14x_4 = 0.15 \\ 0.75x_1 + 0.18x_2 - 0.21x_3 - 0.77x_4 = 0.11 \\ 0.28x_1 - 0.17x_2 + 0.39x_3 + 0.48x_4 = 0.12 \\ 1.00x_1 + 3.14x_2 - 0.21x_3 - 1.00x_4 = -0.11 \end{cases}$$

5. $$\begin{cases} 3.01x_1 - 0.14x_2 + 1.00x_3 - 0.15x_4 = 1.00 \\ -1.75x_1 + 1.11x_2 + 0.13x_3 - 0.75x_4 = 0.13 \\ 0.17x_1 - 2.11x_2 + 0.71x_3 - 1.71x_4 = 1.00 \\ 0.21x_1 + 0.21x_2 + 0.35x_3 + 0.33x_4 = 0.17 \end{cases}$$

6. $$\begin{cases} 1.15x_1 + 0.62x_2 - 0.83x_3 + 0.92x_4 = 2.15 \\ 0.82x_1 - 0.54x_2 + 0.43x_3 - 0.25x_4 = 0.62 \\ 0.24x_1 + 1.15x_2 - 0.33x_3 + 1.42x_4 = -0.62 \\ 0.73x_1 - 0.81x_2 + 1.27x_3 - 0.67x_4 = 0.88 \end{cases}$$

7. $$\begin{cases} 2.2x_1 - 3.17x_2 + 1.24x_3 - 0.87x_4 = 0.46 \\ 1.50x_1 + 2.11x_2 - 0.45x_3 + 1.44x_4 = 1.50 \\ 0.86x_1 - 1.44x_2 + 0.62x_3 + 0.28x_4 = -0.12 \\ 0.48x_1 + 1.25x_2 - 0.63x_3 - 0.97x_4 = 0.35 \end{cases}$$

8. $$\begin{cases} 0.64x_1 + 0.72x_2 - 0.83x_3 + 4.20x_4 = 2.23 \\ 0.58x_1 - 0.83x_2 + 1.43x_3 - 0.62x_4 = 1.71 \\ 0.86x_1 + 0.77x_2 - 1.83x_3 + 0.88x_4 = -0.54 \\ 1.32x_1 - 0.52x_2 - 0.65x_3 + 1.22x_4 = 0.65 \end{cases}$$

9. $$\begin{cases} 1.42x_1 + 0.32x_2 - 0.42x_3 + 0.85x_4 = 1.32 \\ 0.63x_1 - 0.43x_2 + 1.27x_3 - 0.58x_4 = -0.44 \\ 0.84x_1 - 2.23x_2 - 0.52x_3 + 0.47x_4 = 0.64 \\ 0.27x_1 + 1.37x_2 + 0.64x_3 - 1.27x_4 = 0.85 \end{cases}$$

10. $$\begin{cases} 0.73x_1 + 1.24x_2 - 0.38x_3 - 1.43x_4 = 0.58 \\ 1.07x_1 - 0.77x_2 + 1.25x_3 + 0.66x_4 = -0.66 \\ 1.56x_1 + 0.66x_2 + 1.44x_3 - 0.87x_4 = 1.24 \\ 0.75x_1 + 1.22x_2 - 0.83x_3 + 0.37x_4 = 0.92 \end{cases}$$

11. $$\begin{cases} 1.32x_1 - 0.83x_2 - 0.44x_3 + 0.62x_4 = 0.68 \\ 0.83x_1 + 0.42x_2 - 0.56x_3 + 0.77x_4 = 1.24 \\ 0.58x_1 - 0.37x_2 + 1.24x_3 - 0.62x_4 = 0.87 \\ 0.35x_1 + 0.66x_2 - 1.38x_3 - 0.93x_4 = -1.08 \end{cases}$$

12. $$\begin{cases} 0.11x_1 - 0.17x_2 + 0.72x_3 - 0.34x_4 = 0.17 \\ 0.81x_1 + 0.12x_2 - 0.91x_3 + 0.17x_4 = 1.00 \\ 0.17x_1 - 0.18x_2 + 1.00x_3 + 0.23x_4 = 0.21 \\ 0.13x_1 + 0.17x_2 - 0.99x_3 + 0.35x_4 = 2.71 \end{cases}$$

13. $$\begin{cases} 0.18x_1 + 2.11x_2 + 0.13x_3 - 0.22x_4 = 0.22 \\ 0.33x_1 - 0.22x_2 - 1.00x_3 + 0.17x_4 = 0.11 \\ -1.00x_1 + 0.11x_2 + 2.00x_3 - 0.45x_4 = 1 \\ 7x_1 - 0.17x_2 - 0.22x_3 + 0.33x_4 = 0.21 \end{cases}$$

14. $$\begin{cases} 2x_1 + 0.05x_2 - 3.01x_3 - 0.11x_4 = 0.21 \\ x_1 - 2x_2 + 3.02x_3 + 0.05x_4 = 0.18 \\ 0.17x_1 + 0.99x_2 - 2x_3 - 0.17x_4 = 0.17 \\ 0.33x_1 - 0.07x_2 + 0.33x_3 + 2x_4 = 0,17 \end{cases}$$

15. $$\begin{cases} 0.17x_1 - 0.13x_2 - 0.11x_3 - 0.12x_4 = 0.22 \\ x_1 - x_2 - 0.13x_3 + 0.13x_4 = 0.11 \\ 0.35x_1 + 0.33x_2 + 0.12x_3 + 0.13x_4 = 0.12 \\ 0.13x_1 + 0.11x_2 - 0.13x_3 - 0.11x_4 = 1 \end{cases}$$

16. $$\begin{cases} 0.11x_1 + 1.13x_2 - 0.17x_3 + 0.18x_4 = 1 \\ 0.13x_1 - 1.17x_2 + 0.18x_3 + 0.14x_4 = 0.13 \\ 0.11x_1 - 1.05x_2 - 0.17x_3 - 0.15x_4 = 0.11 \\ 0.15x_1 - 0.05x_2 + 0.18x_3 - 0.11x_4 = 1 \end{cases}$$

17. $$\begin{cases} x_1 - 0.17x_2 + 0.11x_3 - 0.15x_4 = 0.17 \\ 0.14x_1 + 0.21x_2 - 0.33x_3 + 0.11x_4 = 1 \\ 0.22x_1 + 3.44x_2 - 0.11x_3 + 0.12x_4 = 2 \\ 0.11x_1 + 0.13x_2 + 0.12x_3 + 0.14x_4 = 0.13 \end{cases}$$

18. $$\begin{cases} x_1 + 0.55x_2 - 0.13x_3 + 0.34x_4 = 0.13 \\ 0.13x_1 - 0.17x_2 + 0.33x_3 + 0.17x_4 = 0.11 \\ 0.11x_1 + 0.18x_2 - 0.22x_3 - 0.11x_4 = 1 \\ 0.13x_1 - 0.12x_2 + 0.21x_3 + 0.22x_4 = 0.18 \end{cases}$$

19. $$\begin{cases} x_1 - 0.51x_2 + 0.12x_3 + 0.55x_4 = 0.12 \\ 0.12x_1 + 0.18x_2 - 0.22x_3 - 0.41x_4 = 0.13 \\ 0.22x_1 - 3.01x_2 + 0.31x_3 + 0.58x_4 = 1 \\ x_1 + 0.24x_2 - 3.05x_3 - 0.22x_4 = 3.41 \end{cases}$$

20. $$\begin{cases} 0.13x_1 + 0.22x_2 - 0.14x_3 + 0.15x_4 = 1 \\ 0.22x_1 - 0.31x_2 + 0.42x_3 - 5.10x_4 = 6.01 \\ 0.62x_1 - 0.74x_2 + 0.85x_3 - 0.96x_4 = 0.11 \\ 0.12x_1 + 0.13x_2 + 0.14x_3 + 0.45x_4 = 0.16 \end{cases}$$

21. $$\begin{cases} 2.7x_1 + 3.3x_2 + 1.3x_3 = 2.1 \\ 3.5x_1 - 1.7x_2 + 2.8x_3 = 1.7 \\ 4.1x_1 + 5.8x_2 - 1.7x_3 = 0.8 \end{cases}$$

22. $$\begin{cases} 1.7x_1 + 2.8x_2 + 1.9x_3 = 0.7 \\ 2.1x_1 + 3.4x_2 + 1.8x_3 = 1.1 \\ 4.2x_1 - 1.7x_2 + 1.3x_3 = 2.8 \end{cases}$$

23. $$\begin{cases} 3.1x_1 + 2.8x_2 + 1.9x_3 = 0.2 \\ 1.9x_1 + 3.1x_2 + 2.1x_3 = 2.1 \\ 7.5x_1 + 3.8x_2 + 4.8x_3 = 5.6 \end{cases}$$

24. $$\begin{cases} 9.1x_1 + 5.6x_2 + 7.8x_3 = 9.8 \\ 3.8x_1 + 5.1x_2 + 2.8x_3 = 6.7 \\ 4.1x_1 + 5.7x_2 + 1.2x_3 = 5.8 \end{cases}$$

25. $$\begin{cases} 3.3x_1 + 2.1x_2 + 2.8x_3 = 0.8 \\ 4.1x_1 + 3.7x_2 + 4.8x_3 = 5.7 \\ 2.7x_1 + 1.8x_2 + 1.1x_3 = 3.2 \end{cases}$$

26. $$\begin{cases} 7.6x_1 + 5.8x_2 + 4.7x_3 = 10.1 \\ 3.8x_1 + 4.1x_2 + 2.7x_3 = 9.7 \\ 2.9x_1 + 2.1x_2 + 3.8x_3 = 7.8 \end{cases}$$

27. $$\begin{cases} 3.2x_1 - 2.5x_2 + 3.7x_3 = 6.5 \\ 0.5x_1 + 0.34x_2 + 1.7x_3 = -0.24 \\ 1.6x_1 + 2.3x_2 - 1.5x_3 = 4.3 \end{cases}$$

28. $$\begin{cases} 5.4x_1 - 2.3x_2 + 3.4x_3 = -3.5 \\ 4.2x_1 + 1.7x_2 - 2.3x_3 = 2.7 \\ 3.4x_1 + 2.4x_2 + 7.4x_3 = 1.9 \end{cases}$$

29. $$\begin{cases} 3.6x_1 + 1.8x_2 - 4.7x_3 = 3.8 \\ 2.7x_1 - 3.6x_2 + 1.9x_3 = 0.4 \\ 1.5x_1 + 4.5x_2 + 3.3x_3 = -1.6 \end{cases}$$

30. $$\begin{cases} 5.6x_1 + 2.7x_2 - 1.7x_3 = 1.9 \\ 3.4x_1 - 3.6x_2 - 6.7x_3 = -2.4 \\ 0.8x_1 + 1.3x_2 + 3.7x_3 = 1.2 \end{cases}$$

9

Computational Eigenvalue Problems

Finding the eigenvalues of a square matrix is a problem that arises in a wide variety of scientific and engineering applications. These applications include, for example, problems involving structural vibrations, energy levels in quantum systems, molecular vibrations, and analysis of systems of linear differential equations. The eigenvalue problem is a special case of the nonlinear problem, so the only way to compute eigenvalues is to use iterative methods. This chapter presents two of the most important numerical techniques for solving eigenvalue problems: the power method and the QR method. We restrict our attention to eigenvalues of real matrices, although much of the theory extends to matrices with complex entries. We begin with a short review of basic facts regarding eigenvalue problems.

9.1 Basic Facts concerning Eigenvalue Problems

Common linear algebra background is presented in section 8.1. Here we will consider some specific definitions related to the eigenvalue problem. In general, this problem is formulated as

$$\mathbf{A}\mathbf{z}_n = \lambda_n \mathbf{M}\mathbf{z}_n, \; n=1, \ldots, N,$$

where $\mathbf{A}$ and $\mathbf{M}$ are square N by N matrices and pair $(\lambda_n, \mathbf{z}_n)$ is an eigenvalue and associated eigenvector. In particular, this problem arises in the analysis of structural vibrations. We restrict our attention to a simpler problem (the standard eigenvalue problem):

$$\mathbf{A}\mathbf{z}_n = \lambda_n \mathbf{z}_n, \; n=1, \ldots, N. \tag{9.1}$$

The spectrum of a matrix $\mathbf{A}$ is the set of all N eigenvalues of $\mathbf{A}$, and these eigenvalues are defined as zeros of the characteristic polynomial $\det(\mathbf{A}-\lambda\mathbf{I})$. Eigenvectors $\mathbf{z}_n$ $(n=1, \ldots, N)$ form a system of orthonormal vectors:

$$(\mathbf{z}_n, \mathbf{z}_m) = \begin{cases} 1, n = m \\ 0, n \neq m \end{cases}.$$

In general, a square matrix **A** may have complex eigenvalues. Only matrices with a special property have exclusively real eigenvalues. These matrices are called Hermitian. Let "H" denote the successive application of two operations: complex conjugation and transposition, then a matrix **A** is Hermitian if $\mathbf{A}=\mathbf{A}^{\mathrm{H}}$. In particular, a real symmetric matrix is a special case of a Hermitian matrix.

To be more specific, let us assume the following numbering of eigenvalues:

$$|\lambda_1| \leq |\lambda_2| \leq \ldots \leq |\lambda_{N-1}| \leq |\lambda_N|.$$

By this means, λ_1 is the eigenvalue with smallest magnitude and λ_N is the eigenvalue with largest magnitude.

The Rayleigh quotient for a matrix **A** is

$$\rho(\mathbf{x}) = \frac{(\mathbf{x}, \mathbf{A}\mathbf{x})}{(\mathbf{x}, \mathbf{x})}$$

for any $\mathbf{x}\neq\mathbf{o}$. It plays an important role in the computation of eigenvalues because of its interesting properties, which are (1) $\lambda_1 \leq \rho(\mathbf{x}) \leq \lambda_N$ for all nonzero $\mathbf{x} \in \mathrm{R}^N$; (2) if $\mathbf{x}=\mathbf{z}_n$, then $\rho(\mathbf{x})=\lambda_n$; and (3) on the basis of the previous property, the residual vector for the eigenvalue problem is introduced as

$$\mathbf{r}(\mathbf{x}) = \mathbf{A}\mathbf{x} - \rho(\mathbf{x})\mathbf{x}$$

and the inequality $||\mathbf{r}(\mathbf{x})|| \leq ||\mathbf{A}\mathbf{x} - \alpha\mathbf{x}||$ holds for any α. Therefore, if $\mathbf{x}$ is some approximation to $\mathbf{z}_n$, then $\rho(\mathbf{x})$ is the best approximation to λ_n.

9.2 Localization of Eigenvalues

Often one can obtain information about the location of eigenvalues in the complex plane without much computational effort. One of the easiest methods to use is based on the following theorem.

Theorem 9.1: Localization of eigenvalues (Gerschgorin)

Any eigenvalue of a matrix $\mathbf{A}=\{a_{nm}\}$ lies at least in one of the disks

$$G_n = \left\{ \lambda : |\lambda - a_{nn}| \leq \sum_{m=1, m\neq n}^{N} |a_{nm}| = sum_n \right\}. \tag{9.2}$$

The set G_n defined in expression (9.2) is called the nth Gerschgorin disk of A. When all eigenvalues are real, then Gerschgorin disks are simply intervals. If n disks form the domain G isolated from other disks, then G contains exactly n eigenvalues of a matrix **A**. Hence the union of the Gerschgorin disks contains all eigenvalues of a matrix **A**. It is sometimes possible to get information about the disk that contains a single eigenvalue. This can be realized for matrices with specific properties, namely: if at some n for all $k=1, \ldots, N\,(k\neq n)$, the following inequalities are fulfilled

$$|a_{kk} - a_{nn}| < sum_k + sum_n\,, \tag{9.3}$$

then the Gerschgorin disk $|\lambda - a_{nn}| \leq sum_n$ contains a single eigenvalue.

9.3 Power Method

This is a classical method used mainly in finding the dominant eigenvalue and eigenvector associated with this eigenvalue. It is not a general method, but it might be useful in a number of situations. For example, it is sometimes an appropriate method for large sparse matrices. In addition, the scheme of this method reveals some important aspects of the eigenvalue problem.

The power method is an iterative process to find eigenvectors of **A**. To begin with, let us perform some preliminary analysis. Take some initial vector $\mathbf{x}^{(0)}$ and suppose that all eigenvalues are different in magnitude. One can expand the vector $\mathbf{x}^{(0)}$ in terms of eigenvectors of **A**:

$$\mathbf{x}^{(0)} = \sum_{n=1}^{N} \alpha_n \mathbf{z}_n\,.$$

Multiplying the vector $\mathbf{x}^{(0)}$ repeatedly by the matrix **A**, we obtain

$$\mathbf{y}^{(k)} = \mathbf{A}^k \mathbf{x}^{(0)} = \sum_{n=1}^{N} \alpha_n \mathbf{A}^k \mathbf{z}_n = \sum_{n=1}^{N} \alpha_n \lambda_n^k \mathbf{z}_n = \lambda_N^k \left(\alpha_N \mathbf{z}_N + \sum_{n=1}^{N-1} \left(\frac{\lambda_n}{\lambda_N} \right)^k \alpha_n \mathbf{z}_n \right).$$

When $k \to \infty$, $\mathbf{y}^{(k)}$ approaches a vector that is collinear with $\mathbf{z}_N$. Thus, taking successive powers of **A** yields information about its eigenvalue λ_N. It is easy to find the dominant eigenvalue from this iteration. Taking into account the second property of the Rayleigh quotient, $\rho(\mathbf{y}^{(k)})$ approaches λ_N when $k \to \infty$.

Incorporating these observations, one can suggest the following implementation of the power method:

1) Define an initial vector $\mathbf{x}^{(0)}$,
2) For every $k=0, 1, \ldots$, compute $\mathbf{y}^{(k+1)} = \mathbf{A}\mathbf{x}^{(k)}$,
3) Compute the next approximation as $\mathbf{x}^{(k+1)} = \mathbf{y}^{(k+1)} / ||\mathbf{y}^{(k+1)}||$. (This procedure, called normalization, is used to prevent either machine overflow or underflow.)
4) Conclude the process if at some k, $||\mathbf{x}^{(k+1)} - \mathbf{x}^{(k)}|| \leq \varepsilon_p$. Then $\mathbf{x}^{(k+1)}$ is an approximate eigenvector $\mathbf{z}_N$ and $s_N = \rho(\mathbf{x}^{(k)})$ is an approximate dominant eigenvalue λ_N.

Convergence of the power method

If all $|\lambda_n|$ are different and $(\mathbf{z}_n, \mathbf{x}^{(0)}) \neq 0$, then

$$\lim_{k \to \infty} \mathbf{x}^{(k)} = \mathbf{z}_N\,,$$

and the following estimate holds:

$$\frac{\left\| \mathbf{x}^{(k)} - \mathbf{z}_N \right\|}{\left\| \mathbf{x}^{(0)} - \mathbf{z}_N \right\|} \leq \left| \frac{\lambda_{N-1}}{\lambda_N} \right|^k = C^k\,.$$

Thus, the rate of convergence of the power method depends on the gap between $|\lambda_{N-1}|$ and $|\lambda_N|$.

If we need only the eigenvalue of a symmetric matrix **A**, then we can perform many fewer iterations. Let $\mathbf{x}^{(k)}$ be some approximation to $\mathbf{z}_n$, so $\mathbf{z}_n = \mathbf{x}^{(k)} + \mathbf{e}^{(k)}$ and $||\mathbf{e}^{(k)}|| = \delta$. The Rayleigh quotient for the vector $\mathbf{x}^{(k)}$ is

$$\rho(\mathbf{x}^{(k)}) = \lambda_n(1 - 4(\mathbf{z}_n, \mathbf{e}^{(k)})^2 + O(\delta^3)) = \lambda_n + O(\delta^2).$$

By this means, approximation $\mathbf{x}^{(k)}$ converges linearly to $\mathbf{z}_n$, but $\rho(\mathbf{x}^{(k)})$ converges quadratically to λ_n.

The following implementation of *power* method is used to find the eigenvalue and eigenvector of the given square matrix. The function takes three input arguments; the square matrix, initial guess vector and error tolerance.

Listing 9.1

```
function [x, l] = pwr(A, x0, tol)
% pwr function uses power method to find the eigen vector
% input:   A - input square matrix
%          x0 - initial guess
%          tol - error tolerance
% output: x - eigenvector
%          l - eigenvalue
i = 0;
x = A*x0;
x = x / norm(x);
while (norm(x - x0) > tol)
   i = i + 1;
   x0 = x;
   x = A*x0;
   x = x / norm(x);
end;
s = sprintf('Total number of iterations = %d', i);
disp(s);
l = (x' * A * x) / (x' * x);
```

We want to find out an eigenvector and the corresponding eigenvalue for the given matrix in the following.

$$\begin{pmatrix} 2 & -1 & 0 & 0 \\ -1 & 2 & -1 & 0 \\ 0 & -1 & 2 & -1 \\ 0 & 0 & -1 & 2 \end{pmatrix}$$

The *pwr* function will be called as follows.

Listing 9.2

```
>> A = [2 -1 0 0; -1 2 -1 0; 0 -1 2 -1; 0 0 -1 2];
>> x0 = [1 0 0 0]';
>> [x, l] = pwr(A, x0, 1e-5)
Total number of iterations = 34
x =
   0.3718
  -0.6015
   0.6015
  -0.3717
l =
   3.6180
```

Vector x is the eigenvector whereas value of l (3.6180) is an eigenvalue of the matrix A. It took 34 iterations to reach the solution with the desired accuracy of 10^{-5}.

9.4 Inverse Iteration

The method of inverse iteration amounts to the power method applied to an appropriate inverse matrix. Let us consider briefly the main ideas behind the method. By multiplying by $\mathbf{A}^{-1}$, we can rewrite eigenvalue problem (9.1) in the following form:

$$\mathbf{A}^{-1}\mathbf{z}_n = \frac{1}{\lambda_n}\mathbf{z}_n, \ \det(\mathbf{A})\neq 0.$$

This means that the eigenvector of a matrix $\mathbf{A}$ is the eigenvector of the matrix $\mathbf{A}^{-1}$ and $\lambda_n(\mathbf{A}^{-1})=1/\lambda_n(\mathbf{A})$. Let us take some initial vector $\mathbf{x}^{(0)}$ and expand it in terms of eigenvectors of $\mathbf{A}$:

$$\mathbf{x}^{(0)} = \sum_{n=1}^{N}\alpha_n\mathbf{z}_n \,.$$

Multiplying the vector $\mathbf{x}^{(0)}$ repeatedly by the matrix $\mathbf{A}^{-1}$, we obtain

$$\mathbf{y}^{(k)} = \left(\mathbf{A}^{-1}\right)^k\mathbf{x}^{(0)} = \sum_{n=1}^{N}\alpha_n\left(\mathbf{A}^{-1}\right)^k\mathbf{z}_n = \sum_{n=1}^{N}\alpha_n\left(\frac{1}{\lambda_n}\right)\mathbf{z}_n =$$

$$\left(\frac{1}{\lambda_1}\right)^k\left(\alpha_1\mathbf{z}_1 + \sum_{n=2}^{N}\left(\frac{\lambda_1}{\lambda_n}\right)^k\alpha_n\mathbf{z}_n\right)$$

When $k\to\infty$, $\mathbf{y}^{(k)}$ approaches a vector that is collinear with $\mathbf{z}_1$. Thus, taking successive powers of $\mathbf{A}^{-1}$ yields information about the eigenvalue of $\mathbf{A}$, which is smallest in magnitude. In practice, we do not compute $\mathbf{A}^{-1}$ explicitly. Instead, it is much less expensive to solve $\mathbf{Ay}=\mathbf{x}$ with respect to $\mathbf{y}$, then compute $\mathbf{y}=\mathbf{A}^{-1}\mathbf{x}$. Incorporating these observations, we can suggest the following implementation of the method of inverse iteration:

(1) Define an initial vector $\mathbf{x}^{(0)}$.
(2) For every k=0, 1, ..., solve the system of linear equations $\mathbf{A}\mathbf{y}^{(k+1)}=\mathbf{x}^{(k)}$.
(3) Compute the next approximation as $\mathbf{x}^{(k+1)}=\mathbf{y}^{(k+1)}/||\mathbf{y}^{(k+1)}||$.
(4) Conclude the process if at some k, $||\mathbf{x}^{(k+1)}-\mathbf{x}^{(k)}||\le\varepsilon_p$. Then $\mathbf{x}^{(k+1)}$ is an approximate eigenvector $\mathbf{z}_1$ and $s_1=\rho(\mathbf{x}^{(k+1)})$ is an approximate eigenvalue λ_1.

Convergence of the method of inverse iteration

If all $|\lambda_n|$ are different and $(\mathbf{z}_1,\mathbf{x}^{(0)})\neq 0$, then

$$\lim_{k\to\infty} \mathbf{x}^{(k)} = \mathbf{z}_1\,,$$

and the following estimate holds:

$$\frac{\left\|\mathbf{x}^{(k)} - \mathbf{z}_1\right\|}{\left\|\mathbf{x}^{(0)} - \mathbf{z}_1\right\|} \le \left|\frac{\lambda_1}{\lambda_2}\right|^k = C^k\,.$$

Thus, the rate of convergence of the method of inverse iteration depends on the gap between $|\lambda_1|$ and $|\lambda_2|$.

The method of inverse iteration is more expensive than the power method, because we have to solve a system of linear equations at every step of the iterative process. However, the matrix **A** remains the same as we iterate, so we can initiate the algorithm by computing a factorization for **A** once (see 8.6), and then each iteration requires only forward and backward substitutions.

The following function implements the *inverse iteration* method. It uses the *Jacobi* function to solve the system of linear equations. The function signature is the same as in previous implementation of *pwr* method.

Listing 9.3

```
function [x, l] = inviter(A, x0, tol)
% inviter   function uses inverse iteration method to find the eigen
% vector and eigen value
% input:  A - input square matrix
%         x0 - initial guess
%         tol - error tolerance
% output: x - eigen vector
%         l - eigen value
i = 0;
y = jacobi(A, x0, tol);
x = y / norm(y);
while (norm(x - x0) > tol)
   i = i + 1;
   x0 = x;
   y = jacobi(A, x0, tol);
   x = y / norm(y);
end;
```

```
s = sprintf('Total number of iterations = %d', i);
disp(s);
l = (x' * A * x) / (x' * x);
```

We apply this function on the same problem as in the last section.

Listing 9.4

```
>> [x, l] = inviter(A, x0, 1e-5)
Total number of iterations = 10
x =
   0.3717
   0.6015
   0.6015
   0.3717
l =
   0.3820
```

This method finds a different eigenvector and eigenvalue compared to last method but it takes only 10 iterations. The actual cost of this method is hidden in solving the system of linear equations using the *Jacobi* method.

9.5 Iteration with a Shift of Origin

Often in practice it is more convenient to apply iterations not to a matrix **A**, but to the matrix $\mathbf{B}=\mathbf{A}-\sigma\mathbf{I}$ (σ=const$\neq$0 is a shift of origin). This essentially does not change eigenvalue problem (9.1), because eigenvectors of **A** coincide with those of the matrix **B** and $\lambda_n(\mathbf{A})=\lambda_n(\mathbf{B})+\sigma$. In this case the power method will converge to some eigenvector $\mathbf{z}_n$ associated with the eigenvalue

$$|\lambda_n(\mathbf{B})| = \max_n |\lambda_n(\mathbf{A}) - \sigma|$$

and the asymptotic error constant is

$$C = \frac{\max\limits_{m\neq n} |\lambda_m(\mathbf{A}) - \sigma|}{\max\limits_{n} |\lambda_n(\mathbf{A}) - \sigma|}.$$

The method of inverse iteration will converge to some eigenvector $\mathbf{z}_n$ associated with the eigenvalue

$$|\lambda_n(\mathbf{B})| = \min_n |\lambda_n(\mathbf{A}) - \sigma|$$

and the asymptotic error constant is

$$C = \frac{\min\limits_{n} |\lambda_n(\mathbf{A}) - \sigma|}{\min\limits_{m\neq n} |\lambda_m(\mathbf{A}) - \sigma|}.$$

There are two situations in which we should use a shift of origin. In the first case, let $|\lambda_N|$ and $|\lambda_{N-1}|$ be equal or very close for a given matrix **A**. By using the power method, we will fail to get convergence either to $\mathbf{z}_N$ or to $\mathbf{z}_{N-1}$ (or convergence will be very slow). When $\sigma \neq 0$, all $\lambda_n(\mathbf{B})$ are different in magnitude; therefore, we will get convergence.

For the second situation in which we should use a shift of origin, suppose we seek the eigenvalue of **A** lying closest to a prescribed point β,

$$|\lambda_m(\mathbf{A}) - \beta| < \min_{n \neq m} |\lambda_n(\mathbf{A}) - \beta|.$$

Then the method of inverse iteration with $\sigma = \beta$ converges to $\mathbf{z}_m$.

In the following, we implement the *power* and *inverse iteration* methods with shift of origin. These implementations are based on the previous implementations. The function signatures remain the same except the addition of shift parameter, s.

Listing 9.5

```
function [x, l] = pwrshft(A, x0, s, tol)
% pwrshft function uses power method with shift to find the
% eigen vector and eigen value
% input:   A - input square matrix
%          x0 - initial guess
%          s - shift factor
%          tol - error tolerance
% output: x - eigen vector
%          l - eigen value
n = length(A);
B = A + s * eye(n);
[x, l] = pwr(B, x0, tol);
l = l - s;
```

Listing 9.6

```
function [x, l] = invitershft(A, x0, s, tol)
% invitershft  function uses inverse iteration method with shift
% to find the eigen vector and eigen value
% input:   A - input square matrix
%          x0 - initial guess
%          s - shift factor
%          tol - error tolerance
% output:  x - eigen vector
%          l - eigen value
n = length(A);
B = A + s * eye(n);
[x, l] = inviter(B, x0, tol);
l = l - s;
```

We apply the above functions to find the eigenvector and eigenvalue for the following matrix.

$$\begin{pmatrix} 1 & 1 & 0 \\ 1 & -1 & 1 \\ 0 & 1 & 1 \end{pmatrix}$$

Listing 9.7

```
>> A = [1 1 0; 1 -1 1; 0 1 1];
>> x0 = [1 0 0]';
>> [x, l] = pwrshft(A, [1 0 0]', 2, 1e-4)
Total number of iterations = 36
x =
  0.6282
  0.4597
  0.6277
l =
  1.7321
>> [x, l] = invitershft(A, [1 0 0]', 2, 1e-4)
Total number of iterations = 5
x =
  0.3251
 -0.8881
  0.3251
l =
 -1.7321
```

9.6 The QR Method

The most widely used method for computing eigenvalues is the QR method. This method has several valuable features: (1) it finds all eigenvalues of a matrix; (2) its behavior in the presence of equal-magnitude eigenvalues allows us to get some useful information about the spectrum of a matrix; and (3) it is simplified when the matrix under consideration is symmetric.

The idea of the QR method is simple. Given **A**, we factorize $\mathbf{A}=\mathbf{A}^{(0)}=\mathbf{QR}$, where **Q** is an orthogonal matrix and **R** is an upper triangular matrix. We then compute a new matrix $\mathbf{A}^{(1)}=\mathbf{RQ}$. The procedure continues by induction: given $\mathbf{A}^{(k)}=\mathbf{Q}^{(k)}\mathbf{R}^{(k)}$, we compute $\mathbf{A}^{(k+1)}=\mathbf{R}^{(k)}\mathbf{Q}^{(k)}$. The procedure is straightforward, but it produces a very interesting result: if all $|\lambda_n|$ are different, then

$$\lim_{k\to\infty} \mathbf{A}^{(k)} = \mathbf{T},$$

where

$$\mathbf{T} = \begin{pmatrix} \lambda_N & * & . & . & * \\ & \lambda_{N-1} & & & . \\ & & . & & . \\ & 0 & & . & * \\ & & & & \lambda_1 \end{pmatrix}. \tag{9.4}$$

By this means, we can compute all eigenvalues in one iterative process. However, the QR method as outlined requires too much arithmetic to be practical. For a general matrix **A**, each QR factorization requires $O(N^3)$ operations. To overcome this obstacle, one can perform an initial transformation on **A** that reduces it to a form for which succeeding QR factorizations are much cheaper. In particular, we begin the QR method by converting **A** to $\mathbf{H}_A$, which is in Hessenberg form:

$$\mathbf{H}_A = \begin{pmatrix} * & . & . & . & * \\ * & . & & & . \\ & . & . & & . \\ & 0 & . & . & . \\ & & & * & * \end{pmatrix}.$$

After this reduction, it is possible to compute the QR factorization of the initial matrix $\mathbf{A}^{(0)}=\mathbf{H}_A$ in $O(N^2)$ operations. More importantly, each matrix $\mathbf{A}^{(k)}$ remains in Hessenberg form. For symmetric matrices, the Hessenberg form is tridiagonal, so much less arithmetic and storage are required. To evaluate convergence, we can monitor the decay of the subdiagonal entries $a_{n+1,n}$ in the matrices $\mathbf{A}^{(k)}$; alternatively we can also monitor the decay of

$$\left| a_{nn}^{(k+1)} - a_{nn}^{(k)} \right|, \; n=1, \ldots, N.$$

The following algorithm summarizes all considerations outlined:

1) Reduce **A** to Hessenberg form, $\mathbf{A} \rightarrow \mathbf{H}_A = \mathbf{A}^{(0)}$.

2) For every $k=0, 1, \ldots$, compute the QR factorization of $\mathbf{A}^{(k)}$: $\mathbf{A}^{(k)}=\mathbf{Q}^{(k)}\mathbf{R}^{(k)}$; then the approximation that follows is $\mathbf{A}^{(k+1)}=\mathbf{R}^{(k)}\mathbf{Q}^{(k)}$.

3) Form two vectors:

$$\mathbf{a}_1^{(k)} = (a_{21}^{(k)}, a_{32}^{(k)}, \ldots, a_{N,N-1}^{(k)})^{\mathrm{T}} \text{ and } \mathbf{a}_2^{(k)} = (a_{11}^{(k)} - a_{11}^{(k-1)}, \ldots, a_{NN}^{(k)} - a_{NN}^{(k-1)})^{\mathrm{T}}.$$

If

$$\left\| \mathbf{a}_1^{(k)} \right\| \le \varepsilon_p \text{ or } \left\| \mathbf{a}_2^{(k)} \right\| \le \varepsilon_p,$$

then the entries $a_{n,n}^{(k)}$ are approximate eigenvalues of **A** and $\left| \lambda_n(\mathbf{A}) - a_{nn}^{(k)} \right| \le \varepsilon_p$.

Convergence of the QR method

If all $|\lambda_n|$ are different, then the matrices $\mathbf{A}^{(k)}$ will converge to an upper triangular matrix **T** (see (9.4)), which contains the eigenvalues λ_n in diagonal positions. If **A** is symmetric, the sequence $\{\mathbf{A}^{(k)}\}$ converges to a diagonal matrix. The following estimate of convergence holds:

$$\left\|\mathbf{T}-\mathbf{A}^{(k)}\right\| \le \left(\max_{1\le n\le N-1}\left|\frac{\lambda_n}{\lambda_{n+1}}\right|\right)^k .$$

As we can see, convergence is absent if only two eigenvalues have equal magnitude (or it is slow, if their magnitudes are close). This is not so rare, especially for nonsymmetric matrices, because in this case eigenvalues occur as complex conjugate pairs. As with the power method of section 9.5, we can use shifts of origin in order to achieve convergence.

The *qr* function to find the QR factors of a matrix is a built-in function so there is no need to implement again. In the following function, we use this method to find out all eigenvalues of a given matrix.

Listing 9.8

```
function l = qrm(A, tol)
% qrm function uses QR method to find the eigen values
% input:   A - input square matrix
%          tol - error tolerance
% output:  l - eigen values
i = 0;
A0 = hess(A);
[Q R] = qr(A0);
A = R * Q;
while (norm(diag(A) - diag(A0)) > tol)
    i = i + 1;
    A0 = A;
    [Q R] = qr(A);
    A = R * Q;
end;
l = diag(A);
s = sprintf('Total number of iterations = %d', i);
disp(s);
```

We want to find out all the eigenvalues of the following matrix.

$$A=\begin{pmatrix}1 & 1 & 0 & 0\\ 1 & -1 & 1 & 0\\ 0 & 0 & 2 & 1\\ 0 & 0 & 1 & -2\end{pmatrix}$$

Listing 9.9

```
>> A = [1 1 0 0; 1 -1 1 0; 0 0 2 1; 0 0 1 -2];
>> l = qrm(A, 1e-6)
Total number of iterations = 0
l =
   1.0000
  -1.0000
   2.0000
  -2.0000
```

The answer is incorrect since it takes zero iteration to reach the solution which is impossible. Now, we will shift the origin of given matrix A by $s = 1$ then apply the same function.

Listing 9.10

```
>> s = 1;
>> B = A + s * eye(length(A))
B =
    2   1   0   0
    1   0   1   0
    0   0   3   1
    0   0   1  -1
>> l = qrm(B, 1e-6) - s
Total number of iterations = 7
l =
   1.4142
  -1.4142
   2.2361
  -2.2361
```

The shift matrix converged to the solution in 7 iterations. In order to get the eigenvalues of original matrix **A**, we subtracted the shift value $s = 1$ from vector l.

9.7 Exercises

Given symmetrical matrix **A**, solve one of the following problems:

(1) calculate the dominant eigenvalue using the power method,
(2) calculate the smallest in magnitude eigenvalue using the inverse iteration,
(3) calculate eigenvalue that is closest to the number

$$\beta = \frac{1}{N}\sum_{n=1}^{N} a_{nn} ,$$

(4) calculate all eigenvalues using the QR method.

Methodical comments: to solve systems of linear equations in the method of inverse iteration use the LU-factorization; to transform matrix into Hessenberg form in QR method use the function `hess(A)`.
Set of symmetrical matrices:

1. $\mathbf{A} = \begin{pmatrix} 1 & 1.5 & 2.5 & 3.5 \\ 1.5 & 1 & 2 & 1.6 \\ 2.5 & 2 & 1 & 1.7 \\ 3.5 & 1.6 & 1.7 & 1 \end{pmatrix}$

2. $\mathbf{A} = \begin{pmatrix} 1 & 1.2 & 2 & 0.5 \\ 1.2 & 1 & 0.4 & 1.2 \\ 2 & 0.4 & 2 & 1.5 \\ 0.5 & 1.2 & 1.5 & 1 \end{pmatrix}$

3. $\mathbf{A} = \begin{pmatrix} 1 & 1.2 & 2 & 0.5 \\ 1.2 & 1 & 0.5 & 1 \\ 2 & 0.5 & 2 & 1.5 \\ 0.5 & 1 & 1.5 & 0.5 \end{pmatrix}$

4. $\mathbf{A} = \begin{pmatrix} 2.5 & 1 & -0.5 & 2 \\ 1 & 2 & 1.2 & 0.4 \\ -0.5 & 1.2 & -1 & 1.5 \\ 2 & 0.4 & 1.5 & 1 \end{pmatrix}$

5. $\mathbf{A} = \begin{pmatrix} 2 & 1 & 1.4 & 0.5 \\ 1 & 1 & 0.5 & 1 \\ 1.4 & 0.5 & 2 & 1.2 \\ 0.5 & 1 & 1.2 & 0.5 \end{pmatrix}$

6. $$\mathbf{A} = \begin{pmatrix} 2 & 1.2 & -1 & 1 \\ 1.2 & 0.5 & 2 & -1 \\ -1 & 2 & -1.5 & 0.2 \\ 1 & -1 & 0.2 & 1.5 \end{pmatrix}$$

7. $$\mathbf{A} = \begin{pmatrix} 2 & 1.5 & 3.5 & 4.5 \\ 1.5 & 2 & 2 & 1.6 \\ 3.5 & 2 & 2 & 1.7 \\ 4.5 & 1.6 & 1.7 & 2 \end{pmatrix}$$

8. $$\mathbf{A} = \begin{pmatrix} 1 & 0.5 & 1.2 & -1 \\ 0.5 & 2 & -0.5 & 0 \\ 1.2 & -0.5 & -1 & 1.4 \\ -1 & 0 & 1.4 & 1 \end{pmatrix}$$

9. $$\mathbf{A} = \begin{pmatrix} 1.2 & 0.5 & 2 & 1 \\ 0.5 & 1 & 0.8 & 2 \\ 2 & 0.8 & 1 & 1 \\ 1 & 2 & 1 & 2 \end{pmatrix}$$

10. $$\mathbf{A} = \begin{pmatrix} 0.5 & 1.2 & 1 & 0.5 \\ 1.2 & 2 & 0.5 & 1.2 \\ 1 & 0.5 & 1 & 1 \\ 0.5 & 1.2 & 1 & 2.2 \end{pmatrix}$$

11. $\mathbf{A} = \begin{pmatrix} 1.2 & 0.5 & 2 & 1 \\ 0.5 & 1 & 0.6 & 2 \\ 2 & 0.6 & 1 & 1 \\ 1 & 2 & 1 & 1 \end{pmatrix}$

12. $\mathbf{A} = \begin{pmatrix} 3 & 1.5 & 4.5 & 5.5 \\ 1.5 & 3 & 2 & 1.6 \\ 4.5 & 2 & 3 & 1.7 \\ 5.5 & 1.6 & 1.7 & 3 \end{pmatrix}$

13. $\mathbf{A} = \begin{pmatrix} 1.6 & 1 & 1.4 & 1 \\ 1 & 1 & 0.5 & 2 \\ 1.4 & 0.5 & 2 & 1.2 \\ 1 & 2 & 1.2 & 0.5 \end{pmatrix}$

14. $\mathbf{A} = \begin{pmatrix} 2.4 & 0.5 & 2 & 1 \\ 0.5 & 1 & 0.8 & 2 \\ 2 & 0.8 & 1 & 0.5 \\ 1 & 2 & 0.5 & 1.2 \end{pmatrix}$

15. $\mathbf{A} = \begin{pmatrix} 0.5 & 1.2 & 2 & 1 \\ 1.2 & 2 & 0.5 & 1.2 \\ 2 & 0.5 & 1 & 0.5 \\ 1 & 1.2 & 0.5 & 1.6 \end{pmatrix}$

16. $$\mathbf{A} = \begin{pmatrix} 1.8 & 1.6 & 1.7 & 1.8 \\ 1.6 & 2.8 & 1.5 & 1.3 \\ 1.7 & 1.5 & 3.8 & 1.4 \\ 1.8 & 1.3 & 1.4 & 4.8 \end{pmatrix}$$

17. $$\mathbf{A} = \begin{pmatrix} 1 & 1.5 & 1.2 & 0.5 \\ 1.5 & 2 & 0.4 & 2 \\ 1.2 & 0.4 & 1.5 & 1.4 \\ 0.5 & 2 & 1.4 & 1.3 \end{pmatrix}$$

18. $$\mathbf{A} = \begin{pmatrix} 1 & 0.5 & -0.5 & 1 \\ 0.5 & -1 & 2 & 0 \\ -0.5 & 2 & 1 & -1.5 \\ 1 & 0 & -1.5 & 2 \end{pmatrix}$$

19. $$\mathbf{A} = \begin{pmatrix} 1 & 1.5 & 0.4 & 2 \\ 1.5 & -1.2 & 1 & -0.5 \\ 0.4 & 1 & 2 & 1.2 \\ 2 & -0.5 & 1.2 & 2.5 \end{pmatrix}$$

20. $$\mathbf{A} = \begin{pmatrix} 1.9 & 1.6 & 1.7 & 1.8 \\ 1.6 & 2.9 & 1.6 & 1.3 \\ 1.7 & 1.6 & 3.9 & 1.4 \\ 1.8 & 1.3 & 1.4 & 4.9 \end{pmatrix}$$

10

Introduction to Finite Difference Schemes for Ordinary Differential Equations

Consideration of problems in the natural sciences and engineering often leads to differential equations, and the study of such problems requires solving these equations. In some cases, it is possible to write their solutions in terms of known elementary functions. As a rule, however, this in principle is impossible, so that the construction of a solution in terms of an explicit closed-form expression cannot be considered as a standard method for solving differential equations. One cannot say that the analytic approach has lost its value. It remains a necessary instrument for the study of simplified, so called model problems. The study of model problems allows one to draw some conclusions about the nature of the more complicated original problems.

10.1 Elementary Example of a Finite Difference Scheme

For the sake of simplicity, consider an example of a difference scheme for the numerical solution of the equation

$$\frac{du}{dx} + \frac{x}{1+x}u = 0\,,\, 0 \le x \le l, \qquad (10.1)$$
$$u(0) = 1\,.$$

Firstly, we introduce a set of points on the interval $[0,l]$: $0=x_0<x_1<\ldots<x_N=l$. This is an example of the simplest computational grid, and the distance between the points $h_n=x_{n+1}-x_n$ is called the step size. In what follows we shall consider the more simple case of the uniform grid, when $h_n=h=\text{const}$. The approximate solution of equation (10.1) is determined at the grid points: that is, it is a set of values

$$u_n^{(a)} = u^{(a)}(x_n)\,,\, n=0,\, \ldots,\, N.$$

This set of values is often called the grid function. Secondly, we need to replace the derivative by the difference approximation. The simplest way to do this is based on the derivative definition:

$$\frac{du}{dx}(x) \approx \frac{u(x+\Delta x)-u(x)}{\Delta x},$$

and it is permissible if Δx is taken sufficiently small. After introduction of this difference approximation we get, in place of differential equation (10.1), the difference equation at every point of the grid ($x \to x_n$, $\Delta x \to h$):

$$\frac{u_{n+1}^{(a)}-u_n^{(a)}}{h}+\frac{x_n}{1+x_n}u_n^{(a)}=0, \qquad (10.2)$$

$$u_0^{(a)}=1,$$

$$x_n = nh,\ n=0,\ \dots,\ N-1,\ h=\frac{l}{N}.$$

The set of difference equations defined on some grid is called the finite difference scheme (or simply the difference scheme). Difference scheme (10.2) can be used to compute the approximate solution. To implement this computation we rewrite (10.2) in the form of a recursion relation:

$$u_{n+1}^{(a)}=\left(1-\frac{hx_n}{1+x_n}\right)u_n^{(a)}.$$

Starting with

$$u_0^{(a)}=1$$

and sequentially taking $n=0,\ \dots,\ N-1$, we find the approximate solution at all grid points. Generally speaking, there are many ways to construct a difference scheme for a given differential equation. For instance, we can write the following difference schemes for equation (10.1):

$$\frac{u_{n+1}^{(a)}-u_n^{(a)}}{h}+\frac{x_{n+1}}{1+x_{n+1}}u_{n+1}^{(a)}=0, \qquad (10.3)$$

with the recursion form

$$u_{n+1}^{(a)}=\left(\frac{1+x_{n+1}}{1+(1+h)x_{n+1}}\right)u_n^{(a)}$$

or

$$\frac{u_{n+1}^{(a)}-u_{n-1}^{(a)}}{2h}+\frac{x_n}{1+x_n}u_n^{(a)}=0, \qquad (10.4)$$

with the recursion form

$$u_{n+1}^{(a)}=u_{n-1}^{(a)}-\frac{2hx_n}{1+x_n}u_n^{(a)}.$$

Naturally, these schemes have different properties. We can see these differences in Figure 10.1, which shows the results of computations with difference schemes (10.2), (10.3), and (10.4), together with the exact solution of equation (10.1).

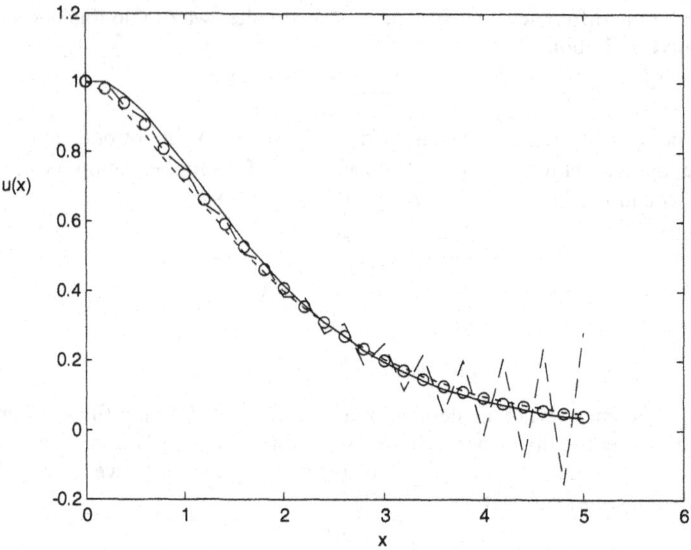

Figure 10.1 Approximate solutions for problem (10.1) generated by different schemes (l=5, h=0.2); open circles, exact solution $u(x)=(1+x)\exp(-x)$; solid line, scheme (10.2); short-dashed line, scheme (10.3); dashed line, scheme (10.4).

In the following program, we solve the problem in equation 10.1 using the strategies given in equations 10.3 and 10.4.

Listing 10.1

```
1  l = 5; h = 0.1; x = 0:h:5; n = length(x);
2  u = zeros(1, n); u(1) = 1;
3  for i=2:n
4       u(i) = ((1+x(i))/(1+(1+h)*x(i))) * u(i-1);
5  end;
6  plot(x, u); hold on;
7  v = zeros(1, n); v(1) = 1; v(2) = 1;
8  for i=2:n-1
9       v(i+1) = v(i-1) - ((2*h*x(i-1)*v(i)) / (1+x(i-1)));
10 end;
11 plot(x, v); hold off;
```

Line 1, sets the length of x i.e. l, defines the step size h and creates the vector x. Line 2, initializes the output vector u and sets the initial value. Lines 3 – 5 implement the recursive form of equation 10.3. Line 7, initializes the output vector for second scheme. Lines 8 – 10 implement the second recursive scheme in equation 10.4. The following output of the gnuplot shows both results and they are same as the plots in figure 10.1.

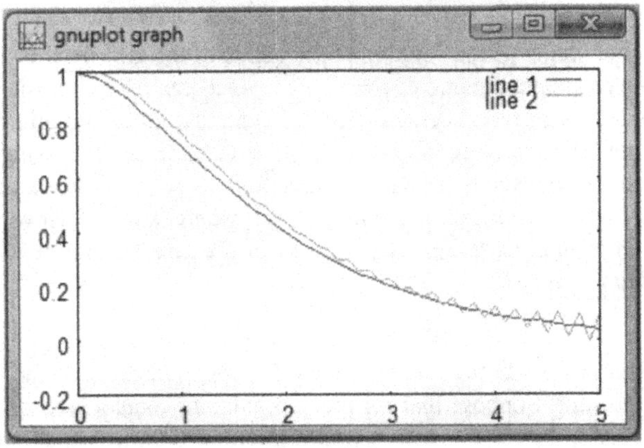

Figure 10.2

10.2 Approximation and Stability

In this section, we consider definitions of approximation, stability, and convergence. Suppose that a differential initial- or boundary value problem is given on some interval I. This means that we want to find the solution $u(x)$ of differential equation (or system of equations) in the interval I under auxiliary conditions on $u(x)$ at one or both ends of this interval. We will write the problem in the symbolic form

$$Du = f\,, \tag{10.5}$$

where D is a given differential operator and f is a given right-hand side. Thus, for example, to write problem (10.1) in form (10.5) we need only take

$$Du \equiv \begin{cases} \dfrac{du}{dx} + \dfrac{x}{1+x}u\,,\, 0 \le x \le l \\ u(0) \end{cases},\; f \equiv \begin{cases} 0 \\ 1 \end{cases}.$$

We will assume that the solution $u(x)$ of problem (10.5) on the interval I exists. In order to calculate this solution by the method of finite differences, we choose a finite set of points on the interval, I. This set is called a grid and will be designated as I_h=$\{x_n\}$ (Figure 10.2).

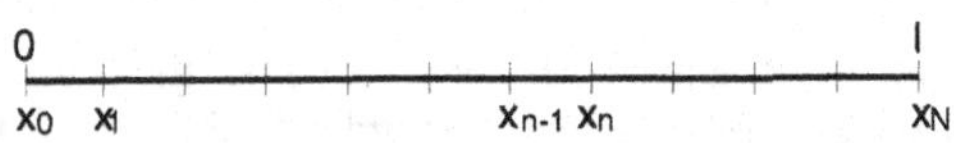

Figure 10.3 One-dimensional grid.

Next, we set out to find not the solution $u(x)$ of problem (10.5), but a grid function

$$\mathbf{u}^{(e)} = \left(u_1^{(e)},\ldots, u_N^{(e)}\right) = (u(x_1),\ldots,u(x_N)),$$

which is a set of values of the solution at the points of the grid, I_h. It is assumed that the grid I_h depends on a parameter $h=x_{n+1}-x_n$, which can take on positive values as small as desired. We are concerned with the computation of the grid function $\mathbf{u}^{(e)}$ because, as the grid is refined, that is as $h\to 0$, it gives us an increasingly complete representation of the solution. Via interpolation it is possible, with increasing accuracy, as $h\to 0$, to construct the solution everywhere within I. However, we will not succeed in computing $\mathbf{u}^{(e)}$ exactly. Instead of the grid function $\mathbf{u}^{(e)}$, we look for another grid function

$$\mathbf{u}^{(a)} = \left(u_1^{(a)},\ldots,u_N^{(a)}\right),$$

which converges to $\mathbf{u}^{(e)}$ as the grid is refined. For this purpose, one can make use of difference equations. Suppose that for the approximate computation of the solution of problem (10.5), we have constructed a difference scheme, which we will write symbolically, by analogy with equation (10.5), in the form

$$D_h\mathbf{u}^{(a)} = \mathbf{f}^{(a)}. \tag{10.6}$$

For example, difference scheme (10.2) may be written in the form (10.6) if we set

$$D_h\mathbf{u}^{(a)} \equiv \begin{cases} \dfrac{u_{n+1}^{(a)} - u_n^{(a)}}{h} + \dfrac{x_n}{1+x_n}u_n^{(a)} \\ u_0^{(a)} \end{cases}, \quad \mathbf{f}^{(a)} \equiv \begin{cases} 0 \\ 1 \end{cases},$$

$$n=0,\ \ldots,\ N-1.$$

Before proceeding further, we need some way of measuring a grid function. In fact, we know this, because grid functions defined on the grid I_h are elements of the linear space, R^N (vectors). Therefore, all definitions considered in section 8.1 are applicable here and the measure of the deviation of $\mathbf{u}^{(a)}$ from $\mathbf{u}^{(e)}$ is taken to be the norm of their difference, that is, the quantity $||\mathbf{u}^{(e)}-\mathbf{u}^{(a)}||$.

When $\mathbf{u}^{(e)}$ is substituted into difference scheme (10.6), some sort of residual will form

$$D_h\mathbf{u}^{(e)} = \mathbf{f}^{(a)} + \delta\mathbf{f}^{(a)}.$$

We will say that the difference scheme $D_h\mathbf{u}^{(a)}=\mathbf{f}^{(a)}$ approximates the problem $Du=f$ on the solution $u(x)$ if $||\delta\mathbf{f}^{(a)}||\to 0$ as $h\to 0$. If, in addition, the inequality

$$\left\|\delta\mathbf{f}^{(a)}\right\| \leq Ch^k$$

is satisfied, where $C>0$ and $k>0$ are constants, then we will say that the approximation (difference scheme) is of the order k with respect to the magnitude of h.

In the above examples, we constructed difference schemes by replacing derivatives in the differential equation with difference expressions. This is a general approach, such that for any differential problem with a smooth enough solution $u(x)$, we can construct a difference scheme with any prescribed order of approximation. Consider examples for constructing some difference expressions that are useful in practice:

$$\frac{du}{dx}(x_n) = \frac{u(x_n+h)-u(x_n)}{h} + O(h) \text{ – the forward difference,}$$

$$\frac{du}{dx}(x_n) = \frac{u(x_n)-u(x_n-h)}{h} + O(h) \text{ – the backward difference,}$$

$$\frac{du}{dx}(x_n) = \frac{u(x_n+h)-u(x_n-h)}{2h} + O(h^2) \text{ – the central difference,}$$

$$\frac{d^2u}{dx^2}(x_n) = \frac{u(x_n+h)-2u(x_n)+u(x_n-h)}{h^2} + O(h^2)\,.$$

Consider the perturbed difference problem

$$D_h\mathbf{z}^{(a)} = \mathbf{f}^{(a)} + \mathbf{e}^{(a)}, \tag{10.7}$$

obtained from problem (10.6) through the addition of a perturbation $\mathbf{e}^{(a)}$ to the right-hand side. We will call difference scheme (10.6) stable if there exist numbers $h_0>0$ and $\delta>0$, such that for any $h<h_0$ and $||\mathbf{e}^{(a)}||<\delta$, difference problem (10.7) has only the solution $\mathbf{z}^{(a)}$. This solution deviates from the solution $\mathbf{u}^{(a)}$ by a grid function $\mathbf{z}^{(a)}-\mathbf{u}^{(a)}$, satisfying the bound $||\mathbf{z}^{(a)}-\mathbf{u}^{(a)}||\leq C||\mathbf{e}^{(a)}||$, where constant C does not depend on h. This inequality means that a small perturbation $\mathbf{e}^{(a)}$ of the right-hand side of difference scheme (10.6) evokes a small perturbation in the solution.

When the differential operator in (10.5) is linear, the following definition of stability is equivalent to that given above. We will call the difference scheme (10.6) stable if for any $\mathbf{f}^{(a)}$ the equation $D_h\mathbf{u}^{(a)}=\mathbf{f}^{(a)}$ has the unique solution $\mathbf{u}^{(a)}$ and

$$\left\|\mathbf{u}^{(a)}\right\| \leq C\left\|\mathbf{f}^{(a)}\right\|, \tag{10.8}$$

where C is some constant not dependent on h.

In practice, we are concerned with how strongly the approximate solution deviates from the exact one and how this deviation can be made sufficiently small.

Suppose that difference scheme (10.6) approximates problem (10.5) on the solution $u(x)$ to order h^k and is stable. Then, the approximate solution $\mathbf{u}^{(a)}$ converges to $\mathbf{u}^{(e)}$, that is, $||\mathbf{u}^{(a)}-\mathbf{u}^{(e)}||\to 0$ as $h\to 0$. In addition, the inequality

$$\left\|\mathbf{u}^{(e)} - \mathbf{u}^{(a)}\right\| \leq Ch^k$$

is satisfied, where C is some constant not depending on h. In this case we say that convergence is of order h^k, or that difference scheme (10.6) has accuracy of kth order. The requirement that a difference scheme be convergent is fundamental. When this requirement is met, then the solution $\mathbf{u}^{(e)}$ can be approximated to any prescribed accuracy, within machine precision, if h is taken sufficiently small.

10.3 Numerical Solution of Initial Value Problems

In the first part of this section we consider various methods for approximating the solution $u(x)$ to a problem of the form

$$\frac{du}{dx} = f(x,u)\,,\ a \le x \le b, \tag{10.9}$$

subject to an initial condition $u(a)=\alpha$.

Later in the section we deal with the extension of those methods to a system of first-order differential equations in the form

$$\frac{d\mathbf{u}}{dx} = \mathbf{f}(x,\mathbf{u}) \tag{10.10}$$

or

$$\frac{du_1}{dx} = f_1(x,u_1,\ldots,u_N)\,,$$
$$\ldots\ldots\ldots\ldots\ldots\ldots\ldots,$$
$$\frac{du_N}{dx} = f_N(x,u_1,\ldots,u_N)\,,$$

for $a \le x \le b$, subject to the initial conditions

$$u_1(a) = \alpha_1\ ,\ \ldots\ ,\ u_N(a) = \alpha_N\,.$$

We assume that the initial value problem (10.10) (or (10.9)) has a unique and stable solution. Solution $\mathbf{u}(x)$ is called stable if for any $\varepsilon>0$ there is a $\delta>0$ such that any other solution $\mathbf{u}^*(x)$ satisfying $||\mathbf{u}(a)-\mathbf{u}^*(a)||\le\delta$ also satisfies $||\mathbf{u}(x)-\mathbf{u}^*(x)||\le\varepsilon$ for all $x\ge a$. Solution $\mathbf{u}(x)$ is called asymptotically stable if, in addition to being stable, it satisfies $||\mathbf{u}(x)-\mathbf{u}^*(x)||\to 0$ as $x\to\infty$. The solution $\mathbf{u}(x)$ is called unperturbed and $\mathbf{u}^*(x)$ perturbed. In other words, stability means that small perturbation in the initial data results in small perturbation in the solution.

Before describing the methods for approximating the solution to problem (10.9), we consider one of the methods for analyzing stability of difference schemes for initial value problems.

10.3.1 Stability of difference schemes for linear problems

In solving initial-value problems, the grid function $\mathbf{u}^{(a)}$ is computed in moving sequentially from one node of the grid to another neighboring node. If we can get a bound on the growth of the solution $\mathbf{u}^{(a)} = \left\{u_n^{(a)}\right\}$ after each such move, we will have at our disposal one of the most widely used methods for the study of stability. We develop this method for a simple test equation,

$$\frac{du}{dx} = -\gamma u\,,\ \gamma>0,\ x\ge 0, \tag{10.11}$$
$$u(0) = \alpha\,.$$

Any difference scheme for this equation may be represented in the following canonical form

$$\mathbf{y}_{n+1} = \mathbf{R}\mathbf{y}_n\,, \tag{10.12}$$
$$n=0,\ \ldots,\ N-1,\ \mathbf{y}_0 \text{ given},$$

where $\mathbf{y}_n$ (generally a vector) depends on $\mathbf{u}^{(a)}$, and $\mathbf{R}=\mathbf{R}(h)$ is the transition operator (generally some matrix). In this case $||\mathbf{u}^{(a)}||=\max||\mathbf{y}_n||$ and $||\mathbf{f}^{(a)}||=||\mathbf{y}_0||$. It is easy to obtain that $\mathbf{y}_n=\mathbf{R}^n\mathbf{y}_0$. Thus we can write

$$\left\|\mathbf{u}^{(a)}\right\| = \max_n \|\mathbf{y}_n\| = \max_n \|\mathbf{R}^n \mathbf{y}_0\| \le \max_n \|\mathbf{R}^n\| \cdot \|\mathbf{y}_0\| \le \max_n \|\mathbf{R}^n\| \cdot \left\|\mathbf{f}^{(a)}\right\|$$

According to our definition of stability (10.8), we get the following condition

$$\max_n \|\mathbf{R}^n\| \le \text{const}$$

or

$$\|\mathbf{R}^n\| \le \|\mathbf{R}\|^n \le \text{const}, \; n=0, \ldots, N-1,$$

which means that the norm of powers of the transition operator is bounded. We certainly satisfy this condition for any N if $||\mathbf{R}||\le 1$. It is known from linear algebra that the spectral radius of a matrix is less than or equal to any norm of the matrix. Taking this into account, we obtain the necessary criterion for stability of a difference scheme for problem (10.11):

$$s(\mathbf{R}) = \max_m |\lambda_m(\mathbf{R})| \le 1, \tag{10.13}$$

where $\lambda_m(\mathbf{R})$ are eigenvalues of the transition operator $\mathbf{R}$.

The stability regions for various schemes, considered below, in the case of test equation (10.11) are given in Table 10.1 The stability condition has the form $\gamma h \le \beta$, so the following table provides values of parameter β. The sign ∞ means unconditional stability.

Table 10.1 Stability regions

Order of approximation, p	1	2	3	4
p-stage explicit Runge-Kutta scheme	2	2	≈2.512	≈2.785
Explicit Adams scheme	2	1	≈0.54	≈0.3
Implicit Adams scheme	∞	∞	≈6	≈3
Predictor-corrector scheme		2	≈1.72	≈1.41

This method is specifically designed for test equation (10.11). However, it can be used for the analysis of difference schemes for equation (10.9). Suppose that the desired integral curve of equation (10.9) passes through the point with coordinates $x=x^*$, $u^*=u(x^*)$. Near this point we have

$$f(x,u) \approx f(x^*,u^*) + \frac{\partial f}{\partial x}(x^*,u^*)(x-x^*) + \frac{\partial f}{\partial u}(x^*,u^*)(u-u^*).$$

Therefore, equation (10.9), to a certain accuracy, may be replaced by equation

$$\frac{du}{dx} = -\gamma^* u + \varphi(x),$$

where

$$\gamma^*(x^*,u^*) = \left|\frac{\partial f}{\partial u}(x^*,u^*)\right| = \text{const},$$

$$\varphi(x) = f(x^*, u^*) + \frac{\partial f}{\partial x}(x^*, u^*)(x - x^*) - u^* \frac{\partial f}{\partial u}(x^*, u^*).$$

Here we assume that differential equation (10.9) is stable, so $\partial f/\partial u$ is nonpositive. This equation appears as our test equation (10.11) (the right-hand function $\varphi(x)$ can be ignored because it has no effect on stability), so the results of stability analysis of difference schemes for that equation may be applied in the general case of equation (10.9). Any conditionally stable scheme applied to test equation (10.11) has a restriction on choosing the step size h: $h \leq \beta/\gamma$, β=const>0. Of course for different points of the integral curve the value of the coefficient $\gamma=\gamma^*$ will differ. Therefore, we should modify this condition, taking into account not only one value of γ^*, but a whole set of such values which sample the range of variations of $\partial f/\partial u$ along the integral curve. A list of two ways to do this follows:

1) we can compute with the variable step size h_n, which satisfies the condition

$$h_n \leq \beta / \gamma_n, \; \gamma_n = \gamma^*(x_n, u_n^{(a)}).$$

2) if we can estimate the value of $u(x)$ on the integration interval, then we can choose the step size from the condition

$$h \leq \frac{\beta}{\max\limits_{a \leq x \leq b} \gamma^*(x, u(x))},$$

and this value of h provides stability of a difference scheme for the whole integration interval.

In the majority of cases encountered in practice, such approaches turn out to be good enough to achieve stability.

10.3.2 Runge-Kutta methods

The simplest Runge-Kutta (RK) scheme for problem (10.9) is one we have already met (see scheme (10.2)). This is the Euler scheme

$$\frac{u_{n+1}^{(a)} - u_n^{(a)}}{h} = f(x_n, u_n^{(a)}), \tag{10.14}$$

$$u_0^{(a)} = \alpha,$$

$$x_n = a + nh, \; n=0, \ldots, N-1, \; h = (b - a)/N,$$

which possesses first-order approximation (and accuracy). First-order schemes provide rather slow convergence, and this results in a considerable amount of computational work in order to achieve the required accuracy. The explicit RK schemes with higher-order approximation are constructed in the following manner:

$$\frac{u_{n+1}^{(a)} - u_n^{(a)}}{h} = b_1 k_1 + \ldots + b_s k_s, \; n=0, \ldots, N-1, \tag{10.15}$$

$$u_0^{(a)} = \alpha,$$

$$k_1 = f(x_n, u_n^{(a)}),$$

$$k_2 = f(x_n + c_2 h, u_n^{(a)} + a_{21} h k_1),$$

$$k_3 = f(x_n + c_3 h, u_n^{(a)} + h(a_{31}k_1 + a_{32}k_2)),$$

$$\dots\dots\dots\dots\dots\dots\dots\dots\dots,$$

$$k_s = f(x_n + c_s h, u_n^{(a)} + h\sum_{m=1}^{s-1} a_{sm}k_m),$$

where s is called the number of stages, and b_l, c_l, a_{lm} (l=1, …, s; m=1, …, s–1) are some real parameters. Upon specifying the number of stages, we can determine these parameters in such a way as to provide the highest possible order of approximation. The Runge-Kutta schemes (10.15) may be symbolically represented in the form of the Butcher table:

$$\begin{array}{c|ccccc}
0 & 0 & & & & \\
c_2 & a_{2,1} & & & & \\
c_3 & a_{3,1} & a_{3,2} & & & \\
\dots & \dots & \dots & \dots & & \\
c_s & a_{s,1} & a_{s,2} & \dots & a_{s,s-1} & \\
\hline
 & b_1 & b_2 & \dots & b_{s-1} & b_s
\end{array}$$

There is a family of RK schemes for each order of approximation. For example, two schemes with the second-order approximation are

$$\begin{array}{c|cc}
0 & & \\
1 & 1 & \\
\hline
 & 1/2 & 1/2
\end{array} \tag{10.16}$$

or

$$\begin{array}{c|cc}
0 & & \\
2/3 & 2/3 & \\
\hline
 & 1/4 & 3/4
\end{array}$$

Next, several higher-order schemes are exemplified. The table

$$\begin{array}{c|ccc}
0 & & & \\
1/2 & 1/2 & & \\
3/4 & 0 & 3/4 & \\
\hline
 & 2/9 & 1/3 & 4/9
\end{array} \tag{10.17}$$

defines the coefficient for the third-order scheme, and tables

$$\begin{array}{c|cccc} 0 & & & & \\ 1/2 & 1/2 & & & \\ 1/2 & 0 & 1/2 & & \\ 1 & 0 & 0 & 1 & \\ \hline & 1/6 & 2/6 & 2/6 & 1/6 \end{array} \tag{10.18}$$

and

$$\begin{array}{c|cccc} 0 & & & & \\ 1/3 & 1/3 & & & \\ 2/3 & -1/3 & 1 & & \\ 1 & 1 & -1 & 1 & \\ \hline & 1/8 & 3/8 & 3/8 & 1/8 \end{array}$$

define coefficients for the fourth-order schemes. The explicit Runge-Kutta schemes are conditionally stable (see Table 10.1). Those results are valid not only for schemes (10.16), (10.17), and (10.18), but also for a variety of other schemes because all p-stage explicit RK schemes of order p have the same stability regions.

The following function implements the fourth order RK scheme. The first argument is a pointer to a function which is the implementation of the function on the right hand-side of equation 10.14. The second argument, *range*, defines the lowest and highest values of independent variable, x. The third argument, *initial_value*, is the initial value of dependent variable, u. The last argument, N, defines the number of steps to solve the problem or the number of partitions of the *range*.

Listing 10.2

```
function [x y] = rk4(f, range, initial_value, N)
% This method computes the rk4 approximation of the
% input function f in the range [a,b] with the initial
% condition y(a) = initial_value and number of steps N
a = range(1); b = range(2); h = (b-a)/N;
x = a:h:b; y = zeros(1, N+1); y(1) = initial_value;
for n=1:length(y)-1
    f1 = h * feval(f, x(n), y(n));
    f2 = h * feval(f, x(n) + h / 2, y(n) + h * f1 / 2);
    f3 = h * feval(f, x(n) + h / 2, y(n) + h * f2 / 2);
    f4 = h * feval(f, x(n) + h, y(n) + h * f3);
    y(n+1) = y(n) + (f1 + 2 * f2 + 2 * f3 + f4) / 6;
end;
x = x';
y = y';
```

Lines 8–11, make four calls to the "*feval*" function. This is a rather expensive function since it dynamically loads the input function and calculates the values for the pointed function. These lines are enclosed in a "*for*" loop which is executed *N* times. This increases the number of calls to "*feval*" by a factor of *N*. This rather clean implementation has a lot of room for improvement in terms of performance. A simple approach is to replace the "*feval*" call to explicit calculation of the function, *f*.

Listing 10.3

```
1 function [x y] = rk4_2(range, initial_value, N)
2 a = range(1); b = range(2); h = (b-a)/N;
3 x = a:h:b; y = zeros(1, N+1); y(1) = initial_value;
4 for n=1:length(y)-1
5     t1 = - (x(n) * y(n))/(1 + x(n));  f1 = h * t1;
6     t2 = - ((x(n)+h/2) * (y(n)+h*f1/2))/(1 + (x(n)+h/2)); f2 = h * t2;
7     t3 = - ((x(n)+h/2) * (y(n)+h*f2/2))/(1 + (x(n)+h/2)); f3 = h * t3;
8     t4 = - ((x(n)+h) * (y(n)+h*f3))/(1 + (x(n)+h));  f4 = h * t4;
9     y(n+1) = y(n) + (f1 + 2 * f2 + 2 * f3 + f4) / 6;
10 end;
11 x = x'; y = y';
```

The *t*'s within the lines 5–8 in the above code listing are the explicitly calculated values of the function *f*, in equation (10.1). The disadvantage is that this function could not be applied to any arbitrary problem, unlike the previous implementation. An even more efficient implementation of RK scheme is shown in the following function.

Listing 10.4

```
public static List jrk4_2(IMatrix range, IMatrix initialValue, IMatrix n) {
double a = range.getMatrixElement(1, 1).getReal();
double b = range.getMatrixElement(1, 2).getReal();
int N = (int) n.getMatrixElement(1).getReal();
double h = (b - a) / N;
double[][] x = new double[N+1][1]; double[][] y = new double[N+1][1];
x[0][0] = a; y[0][0] = initialValue.getMatrixElement(1, 1).getReal();
double f1 = 0; double f2 = 0; double f3 = 0; double f4 = 0;
for (int i=1; i<=N; i++) {
    x[i][0] = x[i-1][0] + h;
    t1 = - (x[i-1][0] * y[i-1][0])/(1 + x[i-1][0]); f1 = h * t1;
    t2 = - ((x[i-1][0]+h/2) * (y[i-1][0]+h*f1/2))/(1 + (x[i-1][0]+h/2)); f2= h*t2;
    t3 = - ((x[i-1][0]+h/2) * (y[i-1][0]+h*f2/2))/(1 + (x[i-1][0]+h/2)); f3=h *t3;
    t4 = - ((x[i-1][0]+h) * (y[i-1][0]+h*f3))/(1 + (x[i-1][0]+h)); f4 = h * t4;
   y[i][0] = y[i-1][0] + (f1 + 2*f2 + 2*f3 + f4) / 6;
}
List rtn = new ArrayList();
```

```
rtn.add(alg.createMatrix(x));
rtn.add(alg.createMatrix(y));
return rtn; }
```

This implementation is the java translation of the previous implementation in m-script, except it is much faster but at the cost of code complexity.

10.3.3 Adams type methods

To obtain u_{n+1} by a Runge-Kutta scheme, with u_n given, one must evaluate the function $f(x,u)$ s times at some intermediate points. The computed s values are then not used any further. In the Adams schemes, to compute the next value u_{n+1} we include not only u_n, but also some approximations prior to u_n. In addition, computation of the value u_{n+1} requires not more than one evaluation of $f(x,u)$, regardless of the order of approximation. The Adams schemes may be obtained as follows. Suppose $u(x)$ is the solution of problem (10.14). If we define $f(x,u(x))=F(x)$, then

$$u(x_{n+1}) - u(x_n) = \int_{x_n}^{x_{n+1}} \frac{du}{dx} dx = \int_{x_n}^{x_{n+1}} F(x)\, dx\,.$$

It is known that there is one and only one polynomial $P_k(x)$ of order not higher than k, which at the points x_n, x_{n-1}, ..., x_{n-k} takes on the values of $F(x_n)$, $F(x_{n-1})$, ..., $F(x_{n-k})$, respectively. This polynomial, for a sufficiently smooth function $F(x)$, deviates from $F(x)$ on the interval $[x_n, x_{n+1}]$ by a quantity of order h^{k+1}, so that

$$\max_x |P_k(x) - F(x)| = O(h^{k+1})\,.$$

The explicit Adams schemes (Adams-Bashforth methods) have the form

$$\frac{u_{n+1}^{(a)} - u_n^{(a)}}{h} = \int_{x_n}^{x_{n+1}} P_k(x)\, dx\,. \qquad (10.19)$$

In general, the explicit Adams schemes may be represented in the following form

$$\frac{u_{n+1}^{(a)} - u_n^{(a)}}{h} = \sum_{k=0}^{p-1} a_k f(x_{n-k}, u_{n-k}^{(a)})\,,\ n=p-1,\ \ldots,\ N-1, \qquad (10.20)$$

where parameters a_k are given in Table 10.2.

Table 10.2 Coefficients of explicit Adams schemes up to order four.

Order of approximation, p	a_k , $k=0$, ..., $p-1$			
1	1			
2	3/2	–1/2		
3	23/12	–16/12	5/12	
4	55/24	–59/24	37/24	–9/24

To start computing via scheme (10.20) when $p \geq 2$, one must know p values of $u_m^{(a)}$, $m=0, \ldots, p-1$, but only $u_0^{(a)} = \alpha$ is given. These values may be found by the Runge-Kutta methods or by expansion of the solution in a Taylor series about the point $x=a$. Let us consider the latter possibility more closely. The value $u(x_m)=u(a+mh)$ may be calculated through the following expansion

$$u(a+mh) = u(a) + mh\frac{du}{dx}(a) + \frac{1}{2!}(mh)^2\frac{d^2u}{dx^2}(a) + \ldots + \frac{1}{p!}(mh)^p\frac{d^pu}{dx^p}(a) + O(h^{p+1})$$

Using equation (10.9) one can express the higher derivatives of $u(x)$ via the function $f(x,u)$:

$$y_k(x,u) = \frac{d^ku}{dx^k} = \frac{dy_{k-1}}{dx} = \frac{\partial y_{k-1}}{\partial x} + \frac{\partial y_{k-1}}{\partial u}\frac{du}{dx} = \frac{\partial y_{k-1}}{\partial x} + f\frac{\partial y_{k-1}}{\partial u},$$

$$y_1(x,u) = f(x,u), \quad k=2, \ldots, p.$$

Thus the starting values $u_m^{(a)}$ are determined as follows:

$$u_m^{(a)} = \alpha + \sum_{k=1}^{p}\frac{(mh)^k}{k!}y_k(a,\alpha), \; m=1,\ldots, p-1.$$

One of the features of the Adams methods is the fact that in the computation of u_{n+1} (given the values of $f(x_{n-1},u_{n-1})$, $f(x_{n-2},u_{n-2})$, ..., already found in the calculation of u_n, u_{n-1}, ...), one needs to compute only the value of $f(x_n,u_n)$. Thus the advantage of the Adams methods over the Runge-Kutta methods consists of the smaller computational effort required for each step. The basic disadvantages are the need for special starting procedures, and the fact that one cannot (without complicating the computational scheme) change the step size h in the course of the computation. The stability regions for explicit Adams schemes in the case of test equation (10.11) are more restrictive than those for the Runge-Kutta methods (see Table 10.1).

If we use the point x_{k+1} to construct polynomial $P_k(x)$ in (10.19), then we obtain the implicit Adams schemes (Adams-Moulton methods). They may be represented in the following form

$$\frac{u_{n+1}^{(a)} - u_n^{(a)}}{h} = \sum_{k=0}^{p-1} b_k f(x_{n+1-k}, u_{n+1-k}^{(a)}), \tag{10.21}$$

$$n = \begin{cases} 0,\ldots,N-1 \text{ for } p=1,2 \\ p-2,\ldots,N-1 \text{ for } p=3,4,\ldots \end{cases},$$

where parameters b_k are given in Table 10.3.

Table 10.3 Coefficients of implicit Adams schemes up to order four

Order of approximation, p	b_k , k=0, ..., p–1			
1	1			
2	1/2	½		
3	5/12	8/12	–1/12	
4	9/24	19/24	–5/24	1/24

The implicit Adams schemes produce more accurate results than the explicit schemes of the same order. In addition, the implicit schemes have wider stability regions. However, the implicit schemes are not computationally efficient, because, in general, difference equation (10.21) cannot be solved explicitly for u_{n+1}. We have to solve a nonlinear equation with respect to u_{n+1} using some iterative technique, but this increases the amount of computational work considerably.

In practice, implicit Adams schemes usually are not used alone. Often, they are used to improve approximations obtained by explicit methods. The combination of an explicit and implicit technique is called a predictor-corrector method. The explicit scheme predicts an approximation, and the implicit scheme corrects this prediction. This procedure can be organized in the following manner:

(1) calculate an approximation $\tilde{u}_{n+1}^{(a)}$ using the explicit Adams scheme of order p as the predictor:

$$\tilde{u}_{n+1}^{(a)} = u_n^{(a)} + h\sum_{k=0}^{p-1} a_k f(x_{n-k}, u_{n-k}^{(a)}) .$$

2) use the implicit Adams scheme of order p as the corrector to improve the approximation:

$$u_{n+1}^{(a)} = u_n^{(a)} + b_0 h f(x_{n+1}, \tilde{u}_{n+1}^{(a)}) + h\sum_{k=1}^{p-1} b_k f(x_{n+1-k}, u_{n+1-k}^{(a)}) .$$

When $p>2$, predictor-corrector methods are still more efficient then RK schemes. These schemes are explicit and they have much wider stability regions in comparison with the explicit Adams methods.

The following function implements the fourth order Adam's scheme. The first argument is a pointer to a function which is the implementation of the function on the right hand-side of equation (10.14). The second argument, *range*, defines the lowest and highest values of independent variable, x. The third argument, *initial_value*, is the initial value of dependent variable, u. The last argument, N, defines the number of steps to solve the problem or the number of partitions of the *range*.

Listing 10.5

```
function [x y] = adams(func, range, initial_value, N)
% This method computes the Adams approximation of the
% input function func in the range [a,b] with the initial
% condition y(a) = initial_value and number of steps N
a = range(1); b = range(2); h = (b-a)/N;
x = a:h:b; y = zeros(1, N+1);
f = zeros(1, N+1); y(1) = initial_value; f(1) = feval(func, x(1), y(1));

y(2) = y(1) + (h/24) * (55*f(1)); %prediction
f(2) = feval(func, x(2), y(2));
y(2) = y(1) + (h/24) * (19*f(1) + 9*f(2)); %correction
f(2) = feval(func, x(2), y(2));
y(3) = y(2) + (h/24) * (-59*f(1) + 55*f(2)); %prediction
f(3) = feval(func, x(3), y(3));
y(3) = y(2) + (h/24) * (-5*f(1) + 19*f(2) + 9*f(3)); %correction
f(3) = feval(func, x(3), y(3));
y(4) = y(3) + (h/24) * (37*f(1) - 59*f(2) + 55*f(3)); %prediction
f(4) = feval(func, x(4), y(4));
y(4) = y(3) + (h/24) * (f(1) - 5*f(2) + 19*f(3) + 9*f(4)); %correction
f(4) = feval(func, x(4), y(4));
for n=4:length(y)-1
   y(n+1) = y(n) + (h/24) * (-9*f(n-3) + 37*f(n-2) - 59*f(n-1) + 55*f(n)); %prediction
   f(n+1) = feval(func, x(n+1), y(n+1));
   y(n+1) = y(n) + (h/24) * (f(n-3) - 5*f(n-2) + 19*f(n-1) + 9*f(n)); %correction
   f(n+1) = feval(func, x(n+1), y(n+1));
end;
x = x'; y = y';
```

The above code is commented to indicate the expressions for prediction and correction. There are a number of calls to *"feval"* function which results in poor performance. As previously mentioned, one solution is to replace these calls with explicit calculation of the function in question or implement the scheme using java language as demonstrated in the case of RK implementation.

10.3.4 Systems of differential equations

In practice, we are frequently faced with an initial value problem, which involves not just a single first-order differential equation but a system of N simultaneous first-order differential equations. All of the above schemes for the numerical solution of the initial-value problem for the first-order differential equation (10.9) automatically generalize to a system of first-order equations (10.10). To see this, we must change $u_n^{(a)}$ to $\mathbf{u}_n^{(a)}$ and $f(x, u_n^{(a)})$ to $\mathbf{f}(x, \mathbf{u}_n^{(a)})$. Then the Runge-Kutta and Adams type schemes preserve their meaning and applicability. For example, the system of equations

$$\begin{aligned} \frac{du_1}{dx} &= (1 - cu_2)u_1\,, x\geq 0,\\ \frac{du_2}{dx} &= (-1 + du_1)u_2\,,\\ u_1(0) &= \alpha_1\,,\quad u_2(0) = \alpha_2\,, \end{aligned} \tag{10.22}$$

may be written in the form

$$\frac{d\mathbf{u}}{dx} = f(x,\mathbf{u})\,,\ x\geq 0,$$
$$\mathbf{u}(0) = \mathbf{u}_{in}\,,$$

if we take

$$\mathbf{u} = \begin{pmatrix} u_1 \\ u_2 \end{pmatrix},$$
$$\mathbf{f}(x,\mathbf{u}) = \begin{pmatrix} (1 - cu_2)u_1 \\ (-1 + du_1)u_2 \end{pmatrix},$$
$$\mathbf{u}_{in} = \begin{pmatrix} \alpha_1 \\ \alpha_2 \end{pmatrix}.$$

For example, the equation for $\mathbf{u}_n^{(a)}$ in Runge-Kutta scheme (10.16),

$$\begin{aligned} \mathbf{u}_{n+1}^{(a)} &= \mathbf{u}_n^{(a)} + \frac{1}{2}h(\mathbf{k}_1 + \mathbf{k}_2)\,,\\ \mathbf{k}_1 &= \mathbf{f}(x_n, \mathbf{u}_n^{(a)})\,,\\ \mathbf{k}_2 &= \mathbf{f}(x_{n+1}, \mathbf{u}_n^{(a)} + h\mathbf{k}_1)\,, \end{aligned}$$

may be written out for equations (10.22) as:

$$\begin{pmatrix} u_{1,n+1}^{(a)} \\ u_{2,n+1}^{(a)} \end{pmatrix} = \begin{pmatrix} u_{1,n}^{(a)} \\ u_{2,n}^{(a)} \end{pmatrix} + \frac{1}{2}h\begin{pmatrix} k_{1,1} + k_{2,1} \\ k_{1,2} + k_{2,2} \end{pmatrix},$$
$$\begin{pmatrix} k_{1,1} \\ k_{1,2} \end{pmatrix} = \begin{pmatrix} (1 - cu_{2,n}^{(a)})u_{1,n}^{(a)} \\ (-1 + du_{1,n}^{(a)})u_{2,n}^{(a)} \end{pmatrix},$$
$$\begin{pmatrix} k_{2,1} \\ k_{2,2} \end{pmatrix} = \begin{pmatrix} (1 - c(u_{2,n}^{(a)} + hk_{1,2}))(u_{1,n}^{(a)} + hk_{1,1}) \\ -1 + d(u_{1,n}^{(a)} + hk_{1,1}))(u_{2,n}^{(a)} + hk_{1,2}) \end{pmatrix}.$$

The initial value problem

$$\frac{d^m u}{dx^m} = f(x, u, \frac{du}{dx}, \ldots, \frac{d^{m-1}u}{dx^{m-1}})\,,\ a\leq x\leq b,$$
$$\frac{d^k u}{dx^k}(a) = \alpha_k\,,\ k=0,\ \ldots,\ m-1,$$

may be reduced to a system of first-order equations (10.10) via changes in the dependent variables. How this can be accomplished is clear from the following example. The equation

$$\frac{d^2u}{dx^2} + \sin(x\frac{du}{du} + u^2) = 0\,,\ x\geq 0,$$

$$u(0) = \alpha_1\,,\ \ \frac{du}{dx}(0) = \alpha_2\,,$$

will take the required form if we set

$$u_1(x) = u(x)\,,\ \ u_2(x) = \frac{du}{dx}\,.$$

We then get

$$\frac{du_1}{dx} = u_2\,,$$

$$\frac{du_2}{dx} = -\sin(xu_2 + u_1^2)\,,$$

$$u_1(0) = \alpha_1\,,\ \ u_2(0) = \alpha_2\,.$$

The following program solves the above system of simultaneous differential equations using the Euler method.

Listing 10.6

```
1  a = 0; b = 5; N = 100;
2  initial_values = [1 0];
3  h = (b-a)/N; x = a:h:b;
4  u1 = zeros(1, N+1); u2 = zeros(1, N+1);
5  u1(1) = initial_values(1); u2(1) = initial_values(2);
6  for n=1:length(x)-1
7          u1(n+1) = u1(n) + h * u2(n);
8          u2(n+1) = u2(n) + h * -sin(x(n)*u2(n)+u1(n).^2);
9  end;
10 plot(x, u2);
```

Line 1 defines the start and end values, $[a,\ b]$, of x and the number of iterations, N. Line 2 sets the initial values α_1 and α_2. Line 3 sets the steps size and the values of x. Line 4 initializes the output vectors u_1 and u_2 to zeros. Line 5 sets their initial values. Lines 6–9 implement the Euler scheme for the simultaneous differential equations. The resultant is plotted in the following figure.

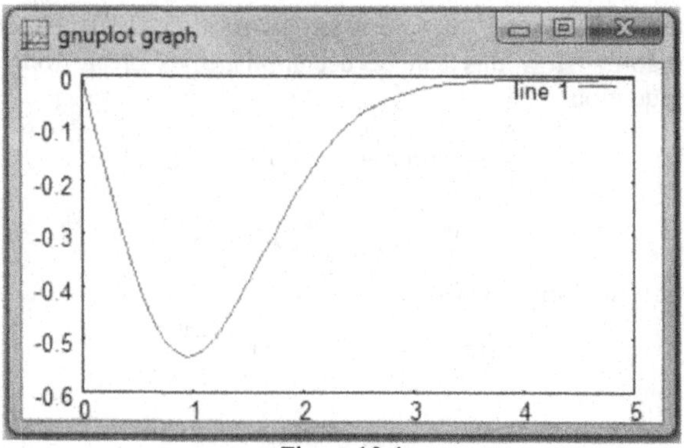

Figure 10.4

10.4 Numerical Solution of Boundary Value Problems

In this section, we show how to approximate the solution to boundary value problems – differential equations where conditions are imposed at different points. There is a great diversity of forms of boundary value problems. We will develop some methods for the numerical solution of such problems using an example of the differential equations of second order, specifically of the form:

$$\frac{d^2u}{dx^2} = f(x, u, \frac{du}{dx}),\ a \leq x \leq b.$$

with the boundary conditions prescribed at $x=a$ and $x=b$. We will assume that $f(\ldots)$ and its partial derivatives with respect to u and du/dx are continuous, and the partial derivative with respect to u is positive, and with respect to du/dx is bounded. These are all reasonable conditions for boundary value problems, which represent physical problems.

10.4.1 Conversion of difference schemes to systems of equations

Here, we consider the suitability of the discretizations studied in section 10.2. To begin with, we look at the linear second-order boundary value problem

$$\frac{d^2u}{dx^2} = p(x)\frac{du}{dx} + q(x)u + r(x),\ a \leq x \leq b, \tag{10.23}$$

$$u(a) = \alpha\,,$$

$$u(b) = \beta\,.$$

Let us introduce a grid with nodes $x_n=a+nh$, $n=0, \ldots, N$, where $h=(b-a)/N$ is the step size. The next step is to substitute, for derivatives in (10.23), the approximations

$$\frac{d^2u}{dx^2}(x_n) \approx \frac{u(x_n+h) - 2u(x_n) + u(x_n-h)}{h^2}, \tag{10.24}$$

$$\frac{du}{dx}(x_n) \approx \frac{u(x_n+h) - u(x_n-h)}{2h}.$$

After performing such substitution, we obtain the following difference scheme:

$$u_0^{(a)} = \alpha,$$

$$\frac{1}{h^2}(u_{n+1}^{(a)} - 2u_n^{(a)} + u_{n-1}^{(a)}) = \frac{p_n}{2h}(u_{n+1}^{(a)} - u_{n-1}^{(a)}) + q_n u_n^{(a)} + r_n, \; n=1, \ldots, N-1,$$

$$u_N^{(a)} = \beta.$$

This difference scheme approximates problem (10.23) to order h^2, because difference approximations for derivatives are of order h^2, and boundary conditions are satisfied exactly. After collecting like terms, our difference scheme may be written as

$$u_0^{(a)} = \alpha,$$

$$(1+\frac{hp_n}{2})u_{n-1}^{(a)} - (2+h^2 q_n)u_n^{(a)} + (1-\frac{hp_n}{2})u_{n+1}^{(a)} = h^2 r_n, \quad (10.25)$$

$$n=1, \ldots, N-1,$$

$$u_N^{(a)} = \beta.$$

It is easy to see that these expressions define a system of linear equations with a tridiagonal matrix (see section 8.6). If $h|p_n|/2<1$ and $q_n>0$, then conditions (8.9) are satisfied and the sweep method provides a stable and efficient technique for solving equations (10.25). Upon solving this system, we obtain $\{u_n\}$, the approximate solution at all grid points.

The construction of a difference scheme for the general nonlinear boundary value problem

$$\frac{d^2u}{dx^2} = f(x, u, \frac{du}{dx}), \; a \leq x \leq b, \quad (10.26)$$

$$u(a) = \alpha,$$

$$u(b) = \beta,$$

is similar to that applied to linear problems. However, the system of equations will not be linear, so some iterative method is required to solve it. If we replace derivatives in (10.26) by their difference approximations (10.24), then the difference scheme may be written as

$$u_0^{(a)} = \alpha,$$

$$\frac{1}{h^2}(u_{n+1}^{(a)} - 2u_n^{(a)} + u_{n-1}^{(a)}) = f\left(x_n, u_n^{(a)}, \frac{u_{n+1}^{(a)} - u_{n-1}^{(a)}}{2h}\right), \quad (10.27)$$

$$n=1, \ldots, N-1,$$

$$u_N^{(a)} = \beta.$$

This scheme can be represented as a system of nonlinear equations of the form

$$\begin{aligned} v_0 &= \alpha\,, \\ v_0 - 2v_1 + v_2 - h^2 f\Big(x_1, v_0, \frac{v_2 - v_0}{2h}\Big) &= 0\,, \\ &\cdots\cdots\cdots\cdots, \\ v_{N-2} - 2v_{N-1} + v_N - h^2 f\Big(x_{N-1}, v_{N-1}, \frac{v_N - v_{N-2}}{2h}\Big) &= 0\,, \\ v_N &= \beta\,. \end{aligned} \tag{10.28}$$

The approximate solution of this system $\mathbf{v}=(v_0, \ldots, v_N)$ approximates $\mathbf{u}^{(a)}$. To solve system (10.28) we can use various iterative methods. The initial approximation $\mathbf{v}^{(0)}$ to $\mathbf{v}$ may be taken as a linear function which passes through the points (a,α) and (b,β):

$$v_n^{(0)} = \frac{\beta - \alpha}{b - a} x_n + \frac{\alpha b - \beta a}{b - a}\,,$$

$$n=0, \ldots, N.$$

The following program solves the above system of simultaneous differential equations using the Euler method.

Listing 10.7

```
11 a = 0; b = 5; N = 100;
12 initial_values = [1 0];
13 h = (b-a)/N; x = a:h:b;
14 u1 = zeros(1, N+1); u2 = zeros(1, N+1);
15 u1(1) = initial_values(1); u2(1) = initial_values(2);
16 for n=1:length(x)-1
17       u1(n+1) = u1(n) + h * u2(n);
18       u2(n+1) = u2(n) + h * -sin(x(n)*u2(n)+u1(n).^2);
19 end;
20 plot(x, u2);
```

Line 1 defines the start and end values, *[a, b]*, of x and the number of iterations, N. Line 2 sets the initial values α_1 and α_2. Line 3 sets the steps size and the values of x. Line 4 initialize the output vectors u_1 and u_2 to zeros. Line 5 sets their initial values. Lines 6–9 implement the Euler scheme for the simultaneous differential equations.

10.4.2 Method of time development

The solution of a boundary value problem may be treated as some equilibrium state. One may consider this equilibrium as the result of the approach to steady state of processes developing in time. Often, the computational treatment of such processes is simpler than the direct calculation of the equilibrium itself. We illustrate the use of the method of time development via the example of an algorithm for the numerical solution of the problem:

$$\frac{d^2u}{dx^2} = c_1(x,u)\frac{du}{dx} + c_2(x)u\,,\ a \le x \le b, \tag{10.29}$$

$$g_1\left(u(a),\frac{du}{dx}(a)\right)=0\,,$$
$$g_2\left(u(b),\frac{du}{dx}(b)\right)=0\,.$$

To begin with, we present some introductory considerations which outline the method of time development. Consider the auxiliary nonstationary problem

$$\frac{\partial v}{\partial t}=\frac{\partial^2 v}{\partial x^2}-c_1(x,v)\frac{\partial v}{\partial x}-c_2(x)v\,,\; a\leq x\leq b,\; t\geq 0, \tag{10.30}$$
$$g_1\left(v(a),\frac{\partial v}{\partial x}(a)\right)=0\,,$$
$$g_2\left(v(b),\frac{\partial v}{\partial x}(b)\right)=0\,,$$
$$v(0,x)=g_0(x)\,,$$

where $g_0(x)$ describes an arbitrary initial condition. Since the boundary conditions are time independent, it is natural to expect that the solution $v(x,t)$ will change more and more slowly with time. Thus, this solution, in the limit as $t\to\infty$, will evolve into the equilibrium solution $u(x)$, characterized by problem (10.29), that is

$$\lim_{t\to\infty} v(x,t)=u(x)\,.$$

Therefore, instead of the stationary problem (10.29) one can solve the nonstationary problem (10.30) until time t, when the solution stops changing within the accuracy we require. This is the idea behind the solution of stationary problems by the method of time development.

In accordance with these considerations we will construct a difference scheme for problem (10.30) instead of (10.29). First, we introduce a computational grid with nodes (x_n,t_k). Now the grid has a little more complex structure than it had before. It is defined by two parameters: $h=x_{n+1}-x_n$, and $\tau_k=t_{k+1}-t_k$, which are called the space step and time step, respectively.

As before, we replace derivatives in (10.30) by their difference approximations, and this results in the following difference scheme:

$$\frac{v_n^{k+1}-v_n^k}{\tau_k}=\frac{v_{n+1}^k-2v_n^k+v_{n-1}^k}{h^2}-c_1(x_n,v_n^k)\frac{v_{n+1}^k-v_{n-1}^k}{2h}-c_2(x_n)v_n^k, \tag{10.31}$$
$$n=1,\,\ldots,\,N-1,\; k=0,\,1,\,\ldots,$$
$$g_1^*(v_0^{k+1},v_1^{k+1})=0\,,$$
$$g_2^*(v_{N-1}^{k+1},v_N^{k+1})=0\,,$$
$$v_n^0=g_0(x_n),$$

here upper superscript k denotes kth time level, not the kth power.

To implement the computation of the approximate solution, we rewrite this difference scheme in the form

$$v_n^{k+1} = \left(1 - \frac{2\tau_k}{h^2} - \tau_k c_2(x_n)\right) v_n^k +$$
$$\left(\frac{\tau_k}{h^2} - \frac{\tau_k c_1(x_n, v_n^k)}{2h}\right) v_{n+1}^k + \left(\frac{\tau_k}{h^2} + \frac{\tau_k c_1(x_n, v_n^k)}{2h}\right) v_{n-1}^k, \tag{10.32}$$
$$n=1, \ldots, N-1, \; k=0, 1, \ldots,$$
$$g_1^*(v_0^{k+1}, v_1^{k+1}) = 0, \quad g_2^*(v_{N-1}^{k+1}, v_N^{k+1}) = 0,$$
$$v_n^0 = g_0(x_n).$$

Since the initial condition $v(x,0)=g_0(x)$ implies that $v_n^0 = g_0(x_n)$ for each $n=0,\ldots, N$, these values can be used in difference equation (10.32) to find the value of $\mathbf{v}^1=\{v(x_n,t_1)\}$. Reapplying this procedure once all the approximations $\mathbf{v}^1$ are known, the values $\mathbf{v}^2$, $\mathbf{v}^3$, ... can be obtained in a similar manner.

It is evident from the previous discussion that difference scheme (10.31) is explicit; as we have seen, explicitness implies conditional stability. If we apply the stability analysis to difference scheme (10.31), we can obtain the following criterion

$$\tau_k \le \min(\tau^{(1)}, \tau_k^{(2)}),$$

where

$$\tau^{(1)} = \frac{2h^2}{4 + h^2 \max\limits_x c_2(x)},$$

$$\tau_k^{(2)} = \frac{2h^2\left(2 + h^2 \max\limits_x c_2(x)\right)}{h^2 \max\limits_n \left(c_1(x_n, v_n^k)\right)^2 + \left(2 + h^2 \max\limits_x c_2(x)\right)^2}.$$

The method of time development is a very reliable technique, because if this method converges, it converges from any initial solution. When the approximate solution $\mathbf{v}^k$ stops changing within the accuracy we require, that is,

$$\frac{\left\|\mathbf{v}^{k+1} - \mathbf{v}^k\right\|}{\left\|\mathbf{v}^{k+1}\right\|} \le \varepsilon_p,$$

the processes of time development is completed.

10.4.3 Accurate approximation of boundary conditions when derivatives are specified at boundary points

So far we considered the boundary condition of the first kind, when the solution itself is specified at boundary points. Many physical problems have boundary conditions involving derivatives specified at boundary points. By way of example let us consider the following problem:

$$\frac{d^2u}{dx^2} = f\left(x, u, \frac{du}{dx}\right), \; a \le x \le b, \tag{10.33}$$
$$\frac{du}{dx}(a) = \alpha,$$
$$u(b) = \beta.$$

The finite-difference approximation of the differential operator is the same as that for (10.27):

$$\frac{1}{h^2}\left(u_{n+1}^{(a)} - 2u_n^{(a)} + u_{n-1}^{(a)}\right) = f\left(x_n, u_n^{(a)}, \frac{u_{n+1}^{(a)} - u_{n-1}^{(a)}}{2h}\right), \qquad (10.34)$$

$$h=1, \ldots, N-1.$$

The obvious approach to replace du/dx at the point a is to use a forward difference approximation

$$\frac{u_1^{(a)} - u_0^{(a)}}{h} = \alpha\,. \qquad (10.35)$$

However, this gives an approximation that has order of only h, whereas difference operator (10.34) has approximation of order h^2. This means that the difference scheme described by (10.34) and (10.35) has approximation of order h only. This circumstance can be eliminated. Let us introduce a fictitious node at $x_{-1}=a-h$. This node allows for constructing a more accurate (central-difference) approximation for the boundary condition $du/dx=\alpha$:

$$\frac{u_1^{(a)} - u_{-1}^{(a)}}{2h} = \alpha\,. \qquad (10.36)$$

To exclude the unknown u_{-1}, we write equation (10.34) for $n=0$:

$$\frac{1}{h^2}\left(u_1^{(a)} - 2u_0^{(a)} + u_{-1}^{(a)}\right) = f\left(x_0, u_0^{(a)}, \frac{u_1^{(a)} - u_{-1}^{(a)}}{2h}\right).$$

After substituting (10.36) into previous expression, we finally get

$$u_1^{(a)} - u_0^{(a)} - \frac{1}{2}h^2 f\left(a, u_0^{(a)}, \alpha\right) = \alpha h\,. \qquad (10.37)$$

To find the approximate solution of problem (10.33) we need to solve system (10.28), except that the first equation of that system should be replaced by the following equation:

$$v_1 - v_0 - \frac{1}{2}h^2 f(a, v_0, \alpha) = \alpha h\,.$$

What we have considered is a simple implementation of the more general approach called the method of fictitious domains. This method is successfully used in various situations where boundary conditions involve derivatives.

10.5 Error Estimation and Control

In this section we will investigate several ways to estimate the error and to control it. By specifying an error tolerance ε_p, we will require either a more accurate and more expensive solution or a less accurate and cheaper one.

The essential idea of the methods described in what follows is to use two different approximate solutions to estimate the error. We begin with error estimation in the case of the initial-value problem. Let us calculate two solutions $u_n^{(a)}$ and $\tilde{u}_n^{(a)}$

at x_n, where $\tilde{u}_n^{(a)}$ denotes the more accurate solution. Then $u_n^{(a)} - \tilde{u}_n^{(a)}$ gives an estimate of the local error

$$e_n = u_n^{(e)} - u_n^{(a)}$$

for the less accurate approximate solution. A pair of Runge-Kutta methods of orders p and $p+1$ (called an embedded pair) may be used to compute these solutions. The essential idea of embedded methods is that such a pair will share stage computations. Thus, an s-stage embedded pair may be represented as

$$\begin{array}{c|ccccc}
0 & 0 & & & & \\
c_2 & a_{2,1} & & & & \\
c_3 & a_{3,1} & a_{3,2} & & & \\
\cdots & \cdots & \cdots & \cdots & & \\
c_s & a_{s,1} & a_{s,2} & \cdots & a_{s,s-1} & \\
\hline
 & b_1^{(1)} & b_2^{(1)} & \cdots & b_{s-1}^{(1)} & b_s^{(1)} \\
 & b_1^{(2)} & b_2^{(2)} & \cdots & b_{s-1}^{(2)} & b_s^{(2)}
\end{array}$$

where sets of parameters $b_n^{(1)}$ and $b_n^{(2)}$ (n=1, ..., s) give methods of orders p and $p+1$, respectively.

Once approximate solutions $u_n^{(a)}$ and $\tilde{u}_n^{(a)}$ have been computed with the step size $h_{n,1}$, we can check if

$$\left|\tilde{u}_n^{(a)} - u_n^{(a)}\right| \le \varepsilon_p \, .$$

If this inequality is not satisfied, then the step size $h_{n,1}$ is rejected and another step size $h_{n,2}$ is selected instead. If the method for finding $u_n^{(a)}$ has order p, then

$$\left|\tilde{u}_n^{(a)} - u_n^{(a)}\right| \approx c \cdot h_n^{p+1} ,$$

so we choose $h_{n,2}$ to satisfy

$$h_{n,2} \le h_{n,1} \left(\frac{\varepsilon_p}{\left|\tilde{u}_n^{(a)} - u_n^{(a)}\right|} \right)^{\frac{1}{1+p}} .$$

This process is repeated until an acceptable step size is found. If the step size is accepted then it can be used as an initial step size $h_{n+1,1}=h_{n,2}$ for the next step.

For predictor-corrector methods, the error estimate can be expressed in terms of the difference between the predictor and the corrector. Let us consider the error estimation when the predictor (Adams-Bashforth scheme) and the corrector (Adams-Moulton scheme) have the same order of approximation. Now, for a method of order p, the local error can be written as a predictor,

$$\overline{e}_{n+1} = u_{n+1}^{(e)} - \overline{u}_{n+1}^{(a)} = \alpha_p c h^p + O\left(h^{p+1}\right),$$

or a corrector,

$$e_{n+1} = u_{n+1}^{(e)} - u_{n+1}^{(a)} = \alpha_c ch^p + O\left(h^{p+1}\right),$$

where α_c and α_p are independent of h. Parameters α_c and α_p are obtained from the approximation analysis, and they are given in Table 10.4.

Table 10.4

Order of approximation, p	α_p	α_c
1	$1/2$	$-1/2$
2	$5/12$	$-1/12$
3	$3/8$	$-1/24$
4	$251/720$	$-19/720$

Thus, from these two formulae we can express the principal error term ch^p:

$$ch^p = \frac{u_{n+1}^{(a)} - \overline{u}_{n+1}^{(a)}}{\alpha_p - \alpha_c} + O\left(h^{p+1}\right).$$

Neglecting terms of degree $p+1$ and above, it is easy to make an estimate of the local error for the corrected (more accurate) values:

$$u_{n+1}^{(e)} - u_{n+1}^{(a)} \approx \frac{\alpha_c}{\alpha_p - \alpha_c}\left(u_{n+1}^{(a)} - \overline{u}_{n+1}^{(a)}\right).$$

By performing local extrapolation, one can obtain a more accurate approximate solution. Indeed, if we write

$$u_{n+1}^{(e)} - \left(u_{n+1}^{(a)} + \alpha_c ch^p\right) = u_{n+1}^{(e)} - \widehat{u}_{n+1}^{(a)} = O\left(h^{p+1}\right),$$

then the approximate solution

$$\widehat{u}_{n+1}^{(a)} = \frac{\alpha_p u_{n+1}^{(a)} - \alpha_c \overline{u}_{n+1}^{(a)}}{\alpha_p - \alpha_c}$$

has the principal error of order h^{p+1}.

In the case of the boundary value problem, we can estimate the global error $\mathbf{e}=\mathbf{u}^{(e)}-\mathbf{u}^{(a)}$. Such an estimate is compared against a specified tolerance or used to select a new grid. We discretize and solve a given boundary value problem on a sequence of grids. The error of the solution on the current grid is estimated and this information is used to decide what the next grid should be, if there is a need for a next grid. The first grid is a guess. The error estimation can be achieved using a process of extrapolation. For second-order boundary value problems, we constructed difference schemes which approximate those problems with the order of h^2. Given a solution $\mathbf{u}^{(a)}$ obtained with step size h, and another one $\mathbf{v}^{(a)}$ obtained with step size $h/2$, we have

$$\mathbf{e}_1 = \mathbf{u}^{(e)} - \mathbf{u}^{(a)} = \mathbf{c}h^2 + O(h^4),$$

$$\mathbf{e}_2 = \mathbf{u}^{(e)} - \mathbf{v}^{(a)} = \mathbf{c}\left(\frac{1}{2}h\right)^2 + O(h^4).$$

So

$$\mathbf{e}_1 \approx \frac{4}{3}\left(\mathbf{v}^{(a)} - \mathbf{u}^{(a)}\right)$$

and

$$\mathbf{e}_2 \approx \frac{1}{3}\left(\mathbf{v}^{(a)} - \mathbf{u}^{(a)}\right).$$

If $||\mathbf{e}_2|| \leq \varepsilon_p$, then solution $\mathbf{v}^{(a)}$ is accepted, and no further grid refinement is needed. One can obtain the extrapolated solution

$$\widehat{\mathbf{u}}^{(a)} = \frac{1}{3}\left(4\mathbf{v}^{(a)} - \mathbf{u}^{(a)}\right),$$

which is a fourth-order accurate solution. However, we do not get a good error estimate for this solution.

10.6 Exercises

Consider an initial value problem for the first order ordinary differential equation

$$u' = f(x,u), \quad 0 \leq x \leq 4, \qquad u(0) = 0.$$

Calculate the approximate solutions of this equation with steps 0.05, 0.02, 0.01 using one of the difference schemes. Show the convergence of approximate solution. Plot the obtained solution. To test the validity of the program, solve the test problem

$$u' = \sin(x)(1-u), \quad 0 \leq x \leq 6, \qquad u(0) = 0,$$

which has the exact solution $u(x)$=1-exp(cos(x)-1).

Suggested difference schemes are

(1) second-order Runge-Kutta scheme,
(2) third-order Runge-Kutta scheme,
(3) fourth-order Runge-Kutta scheme,
(4) second-order explicit Adams scheme,
(5) third-order explicit Adams scheme,
(6) fourth-order explicit Adams scheme,
(7) second-order predictor-corrector method,
(8) third-order predictor-corrector method,
(9) fourth-order predictor-corrector method.

Set of initial value problems:

1. $u' = 1 + 0.2u\sin(x) - u^2, \quad u(0) = 0.$

2. $u' = \cos(x+u) + 0.5(x-u), \quad u(0) = 0.$

3. $u' = \dfrac{\cos(x)}{x+1} - 0.5u^2, \quad u(0) = 0.$

4. $u' = (1 - u^2)\cos(x) + 0.6u, \quad u(0) = 0$.

5. $u' = 1 + 0.4u\sin(x) - 1.5u^2, \quad u(0) = 0$.

6. $u' = \dfrac{\cos(u)}{x+2} + 0.3u^2, \quad u(0) = 0$.

7. $u' = \cos(1.5x + u) + (x - u), \quad u(0) = 0$.

8. $u' = 1 - \sin(x + u) + \dfrac{0.5u}{x+2}, \quad u(0) = 0$.

9. $u' = \dfrac{\cos(u)}{1.5+x} + 0.1u^2, \quad u(0) = 0$.

10. $u' = 0.6\sin(x) - 1.25u^2 + 1, \quad u(0) = 0$.

11. $u' = \cos(2x + u) + 1.5(x - u), \quad u(0) = 0$.

12. $u' = 1 - \dfrac{0.1u}{x+2} - \sin(2x + u), \quad u(0) = 0$.

13. $u' = \dfrac{\cos(u)}{1.25+x} - 0{,}1u^2, \quad u(0) = 0$.

14. $u' = 1 + 0.8u\sin(x) - 2u^2, \quad u(0) = 0$.

15. $u' = \cos(1.5x + u) + 1.5(x - u), \quad u(0) = 0$.

16. $u' = 1 - \sin(2x + u) + \dfrac{0.3u}{x+2}, \quad u(0) = 0$.

17. $u' = \dfrac{\cos(u)}{1.75+x} - 0.5u^2, \quad u(0) = 0$.

18. $u' = 1 + (1 - x)\sin(u) - (2 + x)u, \quad u(0) = 0$.

19. $u' = (0.8 - u^2)\cos(x) + 0.3u, \quad u(0) = 0$.

20. $u' = 1 + 2.2\sin(x) - 1.5u^2, \quad u(0) = 0$.

Consider a boundary value problem for the second order ordinary differential equation

$$u'' + g_1(u', u) = g_2(x), \quad a \le x \le b,$$

$$c_1 u'(a) + c_2 u(a) = \alpha, \quad c_3 u'(b) + c_4 u(b) = \beta.$$

Calculate the approximate solutions of this equation with two steps: h and $h/2$. Estimate the error of the more accurate solution. Plot the obtained solution. Suggested difference schemes are

(1) difference scheme (10.25) (problems 1–20),
(2) method of time development (10.32) (problems 11–20).

Set of boundary value problems:

1. $u'' + \dfrac{u'}{x} + 2u = x$

$$\begin{cases} u(0.1) = 0.5, \\ 2u(1) + 3u'(1) = 1.2. \end{cases}$$

2. $u'' - xu' + 2u = x + 1$

$$\begin{cases} u(0.2) - 0.5u'(0.2) = 2, \\ u(1.2) = 1. \end{cases}$$

3. $u'' + xu' + u = x + 1$

$$\begin{cases} u(0) + 2u'(0) = 1, \\ u'(1) = 1.2. \end{cases}$$

4. $u'' - u' + \dfrac{2u}{x} = x + 0.4$

$$\begin{cases} u(0.1) - 0{,}5u'(0.1) = 2, \\ u'(1.1) = 4. \end{cases}$$

5. $u'' - 3u' + \dfrac{u}{x} = 1$

$$\begin{cases} u(0.2) = 2, \\ u(0.9) + 2u'(0.9) = 0.7. \end{cases}$$

6. $u'' - 0.5u' + 3u = 2x^2$

$$\begin{cases} u(0) + 2u'(0) = 0.6, \\ u(1) = 1. \end{cases}$$

7. $u'' - 0.5xu' + u = 2$

$$\begin{cases} u(0) = 1.2, \\ u(0.7) + 2u'(0.7) = 1.4. \end{cases}$$

8. $u'' + 2x^2u' + u = x$

$$\begin{cases} 2u(0) - u'(0) = 1, \\ u(0.8) = 3. \end{cases}$$

9. $u'' + 2xu' - 2u = 0.6$

$$\begin{cases} u'(0) = 1, \\ 0.4u(2) - u'(2) = 1. \end{cases}$$

10. $u'' - \dfrac{u'}{3} + xu = 2$

$$\begin{cases} u(0.1) = 1.6, \\ 3u(1.1) - 0.5u'(1.1) = 1. \end{cases}$$

11. $u'' + 2u' - \dfrac{u}{x} = 3$

$$\begin{cases} u(0.1) = 2, \\ 0.5u(1.1) - u'(1.1) = 1. \end{cases}$$

12. $u'' + 2u' - xu = x^2$

$$\begin{cases} u'(0) = 0.7, \\ u(1) - 0.5u'(1) = 1. \end{cases}$$

13. $u'' + 3u' - \dfrac{u}{x} = x + 1$

$$\begin{cases} u(0.2) = 1, \\ 2u(1.2) - u'(1.2) = 0.5. \end{cases}$$

14. $u'' + 1.5u' - xu = 0.5$

$$\begin{cases} 2u(0) - u'(0) = 1, \\ u(1) = 3. \end{cases}$$

15. $u'' + 2xu' - u = 0.4$

$$\begin{cases} 2u(0) + u'(0) = 1, \\ u'(0.6) = 2. \end{cases}$$

16. $u'' + \dfrac{2u'}{x} - 3u = 2$

$$\begin{cases} u'(0.1) = 1.5, \\ 2u(1.1) + u'(1.1) = 3. \end{cases}$$

17. $u'' + 2xu' - 2u = 0.6$

$$\begin{cases} u'(0) = 1, \\ 0.4u(2) - u'(2) = 1. \end{cases}$$

18. $u'' + \dfrac{u'}{x} - 0.4u = 2x$

$$\begin{cases} u(0.1) - 0.3u'(0.1) = 0.6, \\ u'(0.9) = 1.7. \end{cases}$$

19. $u'' + 0.8u' - xu = 1.4$

$$\begin{cases} u(0.1) = 0.5, \\ 2u(2.1) + u'(2.1) = 1.7. \end{cases}$$

20. $u'' + 2u' - \dfrac{u}{x} = \dfrac{1}{x}$

$$\begin{cases} 0.5u(0.2) + u'(0.2) = 1, \\ u(1.2) = 0.8. \end{cases}$$

11

Interpolation and Approximation

11.1 Interpolation

For practical use, it is convenient to have an analytical representation of the relationships between variables in a physical problem, and this representation can be approximately reproduced from data given by the problem. The purpose of such a representation might be to determine the values at intermediate points, to approximate an integral or derivative, or simply to represent the phenomena of interest in the form of a smooth or continuous function.

Interpolation refers to the problem of determining a function that exactly represents a collection of data. The most elementary type of interpolation consists of fitting a polynomial to a collection of data points. For numerical purposes, polynomials have the following advantages: they have derivatives and integrals that are themselves polynomials, and their computation is reduced to additions and multiplications only.

11.1.1 Polynomial interpolation

To begin with, consider the problem of determining a polynomial of degree one that passes through the distinct points (x_0, y_0) and (x_1, y_1). The linear polynomial

$$P_1(x) = \frac{x - x_1}{x_0 - x_1} y_0 + \frac{x - x_0}{x_1 - x_0} y_1$$

has the required property: it is easy to verify that $P_1(x_0) = y_0$ and $P_1(x_1) = y_1$. To generalize the concept of linear interpolation to higher-degree polynomials, consider the construction of a polynomial of degree at most N that has the property

$$P_N(x_n) = y_n\,,\ n=0,\ \ldots,\ N, \tag{11.1}$$

where points x_n and y_n are given. This polynomial is unique and may be represented in the form

$$P_N(x) = \sum_{n=0}^{N} y_n l_n(x), \tag{11.2}$$

with the factors

$$l_n(x) = \prod_{\substack{m=0 \\ m \neq n}}^{N} \frac{x - x_m}{x_n - x_m}, \quad n=0, \ldots, N.$$

Polynomial (11.2) is called the Nth Lagrange interpolating polynomial. For an evaluation of the interpolating polynomial at a single point x without explicitly computing the coefficients of the polynomial, the following Neville scheme is very practical. Firstly we define

$$P_n^{(0)}(x_n) = y_n, \; n=0, \ldots, N,$$

then interpolating polynomials are determined according to the recursive relation

$$P_n^{(k)}(x) = \frac{(x - x_n)P_{n+1}^{(k-1)}(x) - (x - x_{n+k})P_n^{(k-1)}(x)}{x_{n+k} - x_n}, \tag{11.3}$$

$$k=1, \ldots, N, \; n=0, \ldots, N-k.$$

and $P_0^{(N)}(x) = P_N(x)$.

Representation (11.2) is very convenient for theoretical investigations because of its simple structure. However, for practical computations it is suitable only for small N. For large N the Lagrange factors $l_n(x)$ become very large and highly oscillatory. There is another representation of the interpolating polynomial that is more practical for computational purposes. We first need to introduce divided differences. Assume that $N+1$ distinct points $x_0, \ldots, x_N$ and $N+1$ values $y_0, \ldots, y_N$ are given. The zeroth divided differences are simply the values of y_n:

$$D_n^{(0)} = y_n, \; n=0, \ldots, N.$$

The divided differences of order k at the point x_n are recursively defined by

$$D_n^{(k)} = \frac{D_{n+1}^{(k-1)} - D_n^{(k-1)}}{x_{n+k} - x_n},$$

$$k=1, \ldots, N, \; n=0, \ldots, N-k.$$

With this notation, the uniquely determined interpolating polynomial with property (11.1) is given by

$$P_N(x) = D_0^{(0)} + \sum_{m=1}^{N} D_0^{(m)} \prod_{n=0}^{m-1} (x - x_n). \tag{11.4}$$

Polynomial (11.4) is called the Newton interpolating polynomial. This polynomial can also be written in a nested form:

$$P_N(x) = \left\{ \ldots \left\{ D_0^{(N)}(x - x_{N-1}) + D_0^{(N-1)} \right\}(x - x_{N-2}) + \ldots + D_0^{(1)} \right\}(x - x_0) + D_0^{(0)}.$$

Thus, the value of the Newton interpolating polynomial at a point x can be calculated using the following algorithm (the Horner scheme):

$$a_N = D_0^{(N)},$$

$$a_k = a_{k+1}(x - x_k) + D_0^{(k)},$$
$$k = N-1, \ldots, 0,$$

and $P_N(x) = a_0$.

As noted earlier, the Lagrange interpolating polynomial is a linear combination of Lagrange coefficients. Before implementing the function to calculate the polynomial, we implement the function to evaluate the Lagrange coefficient in the following. This function takes four arguments. The first argument is the value of x where we want to calculate the Lagrange coefficient. The second argument is a two column matrix with the first column containing the values of x and the second column containing y. The third argument, n, is the order of the coefficient and the last argument is the order of the polynomial.

Listing 11.1

```
function l = lagrange(x, X, n, N)
% lagrange  function finds out the lagrange coefficient for interpolation
% input:  x - value of point at which lagrange coefficient is required
%         X - This two column matrix contains the x's and y's of input data
%         n - order of the coefficient
%         N - order of the polynomial
% output: l - value of lagrange coefficient
product = 1;
for m=1:N
   if m ~= n
       product = product * (x - X(m, 1))/(X(n, 1) - X(m, 1));
   end;
end;
l = product;
```

In the following, we implement the function to evaluate Lagrange polynomial at a given value of x. The function takes similar parameter except we do not need the order of coefficients.

Listing 11.2

```
function p = lagpoly(x, X, N)
% lagpoly  function finds out the lagrange polynomial
% input:    x - value of point to calculate the polynomial value
%           X - This two column matrix contains the x's and y's of input data
%          N - order of the polynomial
% output: p - value of lagrange polynomial at x
%p = X(:, 2) .* lagrange(x, X, 1:N, N);
len = length(x);
p = zeros(len, 1);
for k=1:len
   sum = 0;
```

```
    for n=1:N
        sum = sum + X(n,2)*lagrange(x(k), X, n, N);
    end;
    p(k) = sum;
end;
```

We apply the above function on the following set of data.

x_n	1	2	4	5
y_n	1	2	3	2

We apply the Lagrange polynomial method in the following to determine the curve representing this dataset.

Listing 11.3

```
>> X = [1 1; 2 2; 4 3; 5 2];
>> u = 1:0.1:5;
>> v = lagpoly(u, X, 4);
>> plot(u, v)
>> xlabel('u->');
>> ylabel('y->');
>> title('Lagrange Interpolation');
```

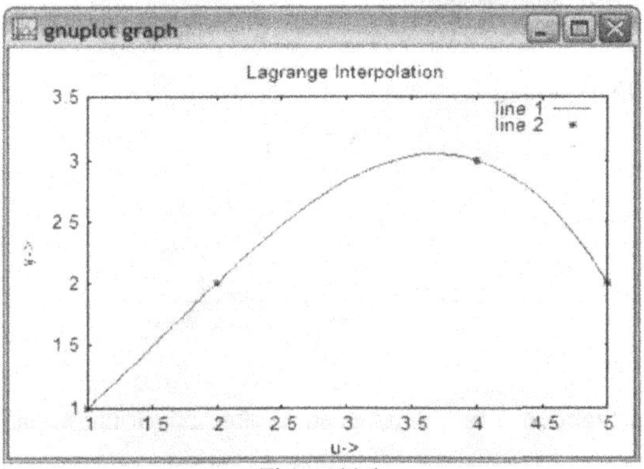

Figure 11.1

11.1.2 Trigonometric interpolation

Periodic functions occur quite frequently in applications. These functions have the property $f(x+T)=f(x)$ for some $T>0$. For example, functions defined on closed planar or spatial curves may always be viewed as periodic functions. Polynomial interpolation is not appropriate for such functions, because algebraic polynomials are not periodic. Therefore, we turn to a consideration of interpolation by trigonometric polynomials. Without loss of generality we assume that the period T is equal to 2π. Assume that $2N+1$ distinct points $x_0, \ldots, x_{2N} \in [0,2\pi)$ and $2N+1$ values $y_0, \ldots, y_{2N}$ are given. Then a uniquely determined trigonometric polynomial $Q_N(x)$ exists and it has the property

$$Q_N(x_n) = y_n, \; n=0, \ldots, 2N.$$

In the Lagrange representation, this polynomial is given by

$$Q_N(x) = \sum_{n=0}^{2N} y_n l_n(x), \tag{11.5}$$

with the factors

$$l_n(x) = \prod_{\substack{m=0 \\ m\neq n}}^{2N} \frac{\sin\left[\frac{1}{2}(x - x_m)\right]}{\sin\left[\frac{1}{2}(x_n - x_m)\right]}, \quad n=0, \ldots, 2N.$$

When the interpolation points are equally spaced, representation (11.5) can be simplified. For an equidistant subdivision with an odd number $2N+1$ of interpolation points

$$x_n = \frac{2\pi n}{2N+1}, \; n=0, \ldots, 2N,$$

there exists a unique trigonometric polynomial

$$Q_N(x) = \frac{a_0}{2} + \sum_{k=1}^{N} (a_n \cos(nx) + b_n \sin(nx)), \tag{11.6}$$

satisfying the interpolation property

$$Q_N(x_n) = y_n, \; n=0, \ldots, 2N.$$

Its coefficients are given by

$$a_n = \frac{2}{2N+1} \sum_{m=0}^{2N} y_m \cos\left(\frac{2\pi mn}{2N+1}\right),$$

$$n=0, \ldots, N,$$

$$b_n = \frac{2}{2N+1} \sum_{m=0}^{2N} y_m \sin\left(\frac{2\pi mn}{2N+1}\right),$$

$$n=1, \ldots, N.$$

For an equidistant subdivision with an even number $2N$ of interpolation points

$$x_n = \frac{\pi n}{N}, \; n=0, \ldots, 2N-1,$$

there exists a unique trigonometric polynomial

$$Q_N(x) = \frac{1}{2}(a_0 + a_N \cos(Nx)) + \sum_{n=1}^{N-1} (a_n \cos(nx) + b_n \sin(nx)) \tag{11.7}$$

satisfying the interpolation property

$$Q_N(x_n) = y_n, \; n=0, \ldots, 2N-1.$$

Its coefficients are given by

$$a_n = \frac{1}{N}\sum_{m=0}^{2N-1} y_m \cos\left(\frac{\pi m n}{N}\right),$$

$$n=0, \ldots, N,$$

$$b_n = \frac{1}{N}\sum_{m=0}^{2N-1} y_m \sin\left(\frac{\pi m n}{N}\right),$$

$$n=1, \ldots, N-1.$$

Trigonometric interpolation polynomials (11.6) and (11.7) may be viewed as the discretized versions of the Fourier series, where the integrals giving the coefficients of the Fourier series are approximated by the rectangular rule at an equidistant partition.

An efficient numerical evaluation of trigonometric polynomials can be done analogously to the Horner scheme for algebraic polynomials. The recursion scheme has the form

$$\begin{aligned} \alpha_{k-1} &= \alpha_k \cos(x) - \beta_k \sin(x) + a_{k-1}, \\ \beta_{k-1} &= \beta_k \cos(x) + \alpha_k \sin(x), \\ & k=N, \ldots, 1, \end{aligned} \tag{11.8}$$

starting with $\alpha_N=a_N$ and $\beta_N=0$. It delivers

$$\alpha_0 = \sum_{n=0}^{N} a_n \cos(nx).$$

If we use b_n instead of a_n and start with $\alpha_N=b_N$ and $\beta_N=0$, then recursion scheme (11.8) delivers

$$\beta_0 = \sum_{n=1}^{N} b_n \sin(nx).$$

Hence the evaluation of a trigonometric polynomial at a point x can be reduced to the evaluation of $\sin(x)$ and $\cos(x)$, and $6N$ additions and $8N$ multiplications.

The trigonometric polynomial is a linear combination of Lagrange like trigonometric coefficients. Before implementing the function to calculate the polynomial, we implement the function to evaluate the trigonometric coefficients in the following. The implementation is very similar to the implementation in the previous section. This function takes four arguments. The first argument is the value of *x* where we want to calculate the coefficient. The second argument is a two column matrix with the first column containing the values of *x* and the second column containing *y*. The third argument, *n*, is the order of the coefficient and the last argument is the order of the polynomial.

Listing 11.4

```
function l = trigcoef(x, X, n, N)
% trigcoef  function finds out the coefficient for trigonometric interpolation
% input:    x - value of point where coefficient value is required
```

```
%           X - This two column matrix contains the x's and y's of input data
%           n - order of the coefficient
%           N - order of the polynomial
% output: l - value of the coefficient
product = 1;
for m=1:N
    if m ~= n
        product = product * sin((x - X(m, 1))/2)/sin((X(n, 1) - X(m, 1))/2);
    end;
end;
l = product;
```

In the following, we implement the function to evaluate the trigonometric polynomial at a given value of x. The function takes a similar parameter except that we do not need the order of coefficients.

Listing 11.5

```
function p = trigonometricpoly(x, X, N)
% trigonometricpoly  function finds out the trigonometric polynomial
% input:   x - value of point where polynomial value is required
%          X - This two column matrix contains the x's and y's of input data
%          N - order of the polynomial
% output:  p - value of trigonometric polynomial at x
len = length(x); p = zeros(len, 1);
for k=1:len
    sum = 0;
    for n=1:N
        sum = sum + X(n,2)*trigcoef(x(k), X, n, N);
    end;
    p(k) = sum;
end;
```

We apply the above function on the following set of data.

x_n	0	0.1667	0.3333	0.5000	0.6667	0.8333	1.0000
y_n	1.0000	-2.000	3.0000	6.0000	0.0000	-1.000	1.0000

We apply the trigonometric polynomial method in the following and determine the curve representing this dataset.

Listing 11.6

```
>> X = [0, 1; 1/6, -2; 1/3, 3; 1/2, 6; 2/3, 0; 5/6, -1; 1, 1];
>> u = 0:0.01:1;
>> v = trignometricpoly(u, X, 7);
```

```
>> plot(u, v)
>> xlabel('x->');
>> ylabel('y->');
>> title('trignometric Interpolation');
>> hold on;
>> plot(X(:,1), X(:,2), '*');
>> hold off;
```

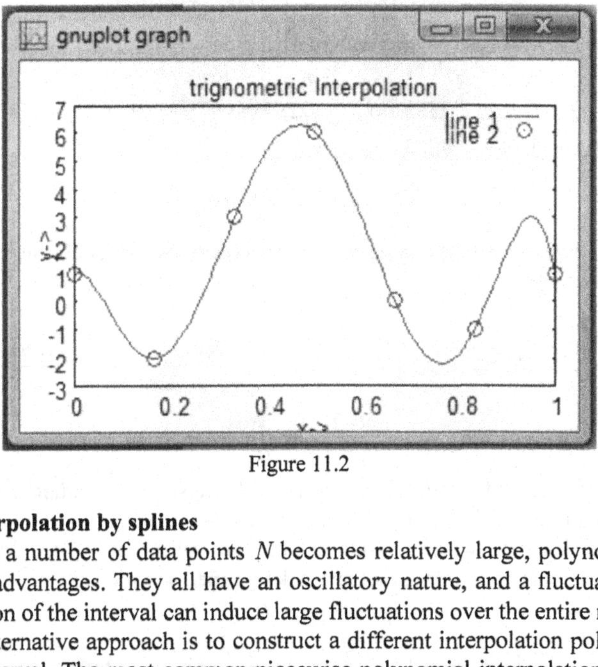

Figure 11.2

11.1.3 Interpolation by splines

When a number of data points N becomes relatively large, polynomials have serious disadvantages. They all have an oscillatory nature, and a fluctuation over a small portion of the interval can induce large fluctuations over the entire range.

An alternative approach is to construct a different interpolation polynomial on each subinterval. The most common piecewise polynomial interpolation uses cubic polynomials between pairs of points. It is called cubic spline interpolation. The term spline originates from the thin wooden or metal strips that were used by draftsmen to fit a smooth curve between specified points. Since the small displacement w of a thin elastic beam is governed by the fourth-order differential equation $d^4w/dx^4=0$, cubic splines indeed model the draftsmen splines.

A cubic spline interpolant $s(x)$ is a combination of cubic polynomials $s_n(x)$ ($s(x)=\cup s_n(x)$), which have the following form on each subinterval $[x_n, x_{n+1}]$:

$$s_n(x) = a_n + b_n(x - x_n) + c_n(x - x_n)^2 + d_n(x - x_n)^3 . \quad (11.9)$$

To determine the coefficients of $s_n(x)$, several conditions are imposed. These include

$$s_n(x_n) = y_n\,,\ n=0,\ \ldots,\ N-1$$

and

$$s_{N-1}(x_N) = y_N \,.$$

Additional conditions follow.

- Continuity of $s(x)$:

$$s_{n-1}(x_n) = s_n(x_n)\,,\ n=1,\ \ldots,\ N-1.$$

- Continuity of the first derivative of $s(x)$:

$$\frac{ds_{n-1}}{dx}(x_n) = \frac{ds_n}{dx}(x_n)\,,\ n=1,\ \ldots,\ N-1.$$

- Continuity of the second derivative of $s(x)$:

$$\frac{d^2s_{n-1}}{dx^2}(x_n) = \frac{d^2s_n}{dx^2}(x_n)\,,\ n=1,\ \ldots,\ N-1.$$

- Some boundary conditions, for example:

$$\frac{d^2s_0}{dx^2}(x_0) = \frac{d^2s_{N-1}}{dx^2}(x_N) = 0\,.$$

These conditions result in the system of linear equations for coefficients c_n of cubic polynomial (11.9):

$$c_0 = 0\,,$$

$$\Delta x_{n-1}c_{n-1} + 2(\Delta x_n + x_{n-1})c_n + \Delta x_n c_n = \frac{3(y_{n+1} - y_n)}{\Delta x_n} - \frac{3(y_n - y_{n-1})}{\Delta x_{n-1}}\,,$$

$$n=1,\ \ldots,\ N-1,$$

$$c_N = 0\,,$$

where $\Delta x_n = x_{n+1} - x_n$. The coefficient matrix of this system is a tridiagonal one, and it also has the property of diagonal dominance. Hence, the solution of this system can be obtained by the sweep method. Once the values of c_n are known, other coefficients of cubic polynomial $s_n(x)$ are determined as follows:

$$a_n = y_n\,,\ n=0,\ \ldots,\ N,$$

$$b_n = \frac{1}{\Delta x_n}(y_{n+1} - y_n) - \frac{\Delta x_n}{3}(c_{n+1} + 2c_n)\,,$$

$$d_n = \frac{1}{3\Delta x_n}(c_{n+1} - c_n)\,,$$

$$n=0,\ \ldots,\ N-1.$$

11.2 Approximation of Functions and Data Representation

Approximation theory involves two types of problems. One arises when a function is given explicitly, but one wishes to find other types of functions which are more convenient to work with. The other problem concerns fitting functions to given data and finding the "best" function in a certain class that can be used to represent the data. We will begin this section with the latter problem.

11.2.1 Least-squares approximation

Consider the problem of estimating the values of a function at nontabulated points, given the experimental data in Table 11.1.

Table 11.1

x_n	0	0.5	1	2	3	4	6	8
y_n	0.08	0.39	0.55	1.02	1.61	2.15	3.0	4.05

Interpolation requires a function that assumes the value of y_n at x_n for each n=1, ..., 11. Figure 11.1 shows a graph of the values in Table 11.1. From this graph, it appears that the actual relationship between x and y is linear. It is evident that no line precisely fits the data because of errors in the data collection procedure. In this case, it is unreasonable to require that the approximating function agrees exactly with the given data. A better approach for a problem of this type would be to find the "best" (in some sense) approximating line, even though it might not agree precisely with the data at any point.

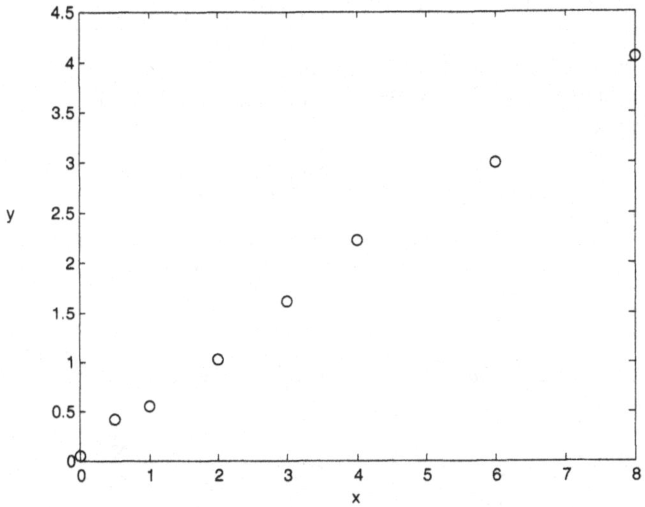

Figure 11.3 Graph of the values listed in Table 11.1.

Let $f_n=ax_n+b$ denote the nth value on the approximating line and y_n the nth given y-value. The general problem of fitting the best line to a collection of data involves minimizing the total error

$$e(a,b) = \sum_{n=1}^{N}(y_n - f_n)^2 = \sum_{n=1}^{N}(y_n - (ax_n + b))^2$$

with respect to the parameters a and b. For a minimum to occur, we need

$$\frac{\partial}{\partial a}\sum_{n=1}^{N}(y_n-(ax_n+b))^2=2\sum_{n=1}^{N}-x_n(y_n-ax_n-b)=0$$

and

$$\frac{\partial}{\partial b}\sum_{n=1}^{N}(y_n-ax_n-b)^2=-2\sum_{n=1}^{N}(y_n-ax_n-b)=0\,.$$

These equations are simplified to the normal equations

$$a\sum_{n=1}^{N}x_n^2+b\sum_{n=1}^{N}x_n=\sum_{n=1}^{N}x_ny_n\,, \qquad (11.10)$$

$$a\sum_{n=1}^{N}x_n+bN=\sum_{n=1}^{N}y_n\,.$$

The solution to this system is as follows:

$$a=\frac{Ng_1-c_2g_2}{Nc_1-c_2^2}\,,\quad b=\frac{c_1g_2-c_2g_1}{Nc_1-c_2^2}\,,$$

where

$$c_1=\sum_{n=1}^{N}x_n^2\,,\quad c_2=\sum_{n=1}^{N}x_n\,,\quad g_1=\sum_{n=1}^{N}x_ny_n\,,\quad g_2=\sum_{n=1}^{N}y_n\,.$$

This approach is called the method of least-squares.

The problem of approximating a set of data (x_n,y_n), $n=1, \ldots, N$ with an algebraic polynomial

$$P_M(x)=\sum_{m=0}^{M}a_mx^m$$

of degree $M<N-1$ using least-squares is resolved in a similar manner. It requires choosing the coefficients $a_0, \ldots, a_M$ to minimize the total error:

$$e(a_0,\ldots,a_M)=\sum_{n=1}^{N}(y_n-P_M(x_n))^2\,.$$

Hence, the conditions $\partial e/\partial a_m=0$ for each $m=0, \ldots, M$ result in a system of $M+1$ linear equations for $M+1$ unknowns a_m,

$$a_0N+a_1\sum_{n=1}^{N}x_n+a_2\sum_{n=1}^{N}x_n^2+\ldots+a_M\sum_{n=1}^{N}x_n^M=\sum_{n=1}^{N}y_n\,,$$

$$a_0\sum_{n=1}^{N}x_n+a_1\sum_{n=1}^{N}x_n^2+\ldots+a_M\sum_{n=1}^{N}x_n^{M+1}=\sum_{n=1}^{N}y_nx_n\,, \qquad (11.11)$$

$$\ldots\ldots\ldots\ldots\ldots\ldots\ldots\ldots\ldots,$$

$$a_0\sum_{n=1}^{N}x_n^M+a_1\sum_{n=1}^{N}x_n^{M+1}+\ldots+a_M\sum_{n=1}^{N}x_n^{2M}=\sum_{n=1}^{N}y_nx_n^M\,.$$

This system will have a unique solution, provided that the x_n are distinct.

We apply the above method on the following set of data.

x_n	0	0.1667	0.3333	0.5000	0.6667	0.8333	1.0000
y_n	1.0000	-2.000	3.0000	6.0000	0.0000	-1.000	1.0000

There are 7 data points ($N = 7$) and we will use a 3rd order polynomial for least square approximation ($M = 3$). We build the following linear system of equations based on the least square method.

$$7.0000a_0 + 3.5000a_1 + 2.5278a_2 + 2.0417a_3 = 8.0000$$
$$3.5000a_0 + 2.5278a_1 + 2.0417a_2 + 1.7554a_3 = 3.8330$$
$$2.5278a_0 + 2.0417a_1 + 1.7554a_2 + 1.5691a_3 = 2.0833$$
$$2.0417a_0 + 1.7554a_1 + 1.5691a_2 + 1.4397a_3 = 1.2731$$

We solve the above system of equations in the following to find out the unknown values of *a*'s.

Listing 11.7

```
>> A = [7.0000    3.5000    2.5278    2.0417;
3.5000    2.5278    2.0417    1.7554;
2.5278    2.0417    1.7554    1.5691;
2.0417    1.7554    1.5691    1.4397];
>> b = [8; 3.833; 2.08333; 1.2731];
>> a = inv(A) * b;
a =
-0.4739       16.7013       -28.8789      12.6672
```

Once we know the polynomial coefficients, we could use the built-in *polyval* function to evaluate the polynomial for different values of the independent variable, x. We use this function in the following to generate the approximate polynomial values and create a plot showing the given data points and the approximating polynomial.

Listing 11.9

```
>> X = [0, 1; 1/6, -2; 1/3, 3; 1/2, 6; 2/3, 0; 5/6, -1; 1, 1];
>> x = X(:,1); y = X(:,2);
>> u = 0:0.01:1;
>> a = [12.6672, -28.8789, 16.7013, -0.4739];
>> v = polyval(a, u);
>> plot(u, v);
>> xlabel('x->');
>> ylabel('y->');
>> title('Least sq. approx.');
```

```
>> hold on;
>> plot(x, y, '*');
>> hold off;
```

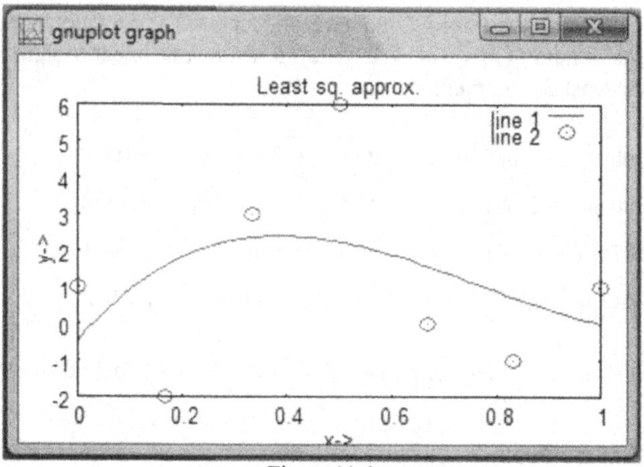

Figure 11.4

The approximating polynomial function nicely goes through the data points. The error could be reduced by increasing the polynomial order. But for the high order polynomial the curve shows oscillatory behavior so we could only use the low order polynomial.

11.2.2 Approximation by orthogonal functions

We now turn our attention to the problem of function approximation. The previous section was concerned with the least-squares approximation of discrete data. We now extend this idea to the polynomial approximation of a given continuous function $f(x)$ defined on an interval $[a,b]$. The summations of previous section are replaced by their continuous counterparts, namely defined integrals, and a system of normal equation is obtained as before. For the least-squares polynomial approximation of degree M, the coefficients in

$$P_M(x) = \sum_{m=0}^{M} a_m x^m$$

are chosen to minimize

$$e(a_0,\ldots,a_M) = \int_a^b (f(x) - P_M(x))^2 \, dx \, .$$

To determine these coefficients we need to solve the following system of linear equations

$$\sum_{m=0}^{M} a_m \int_a^b x^{n+m}\,dx = \int_a^b x^n f(x)\,dx\,, \tag{11.12}$$

$$n=0,\ \ldots,\ M.$$

These equations always possess a unique solution. The (n,m) entry of the coefficient matrix of system (11.12) is

$$\int_a^b x^{n+m}\,dx = \frac{b^{n+m+1} - a^{n+m+1}}{n+m+1}\,.$$

Matrices of this form are known as Hilbert matrices and they are very ill conditioned. Numerical results obtained using these matrices are sensitive to rounding errors and must be considered inaccurate. In addition, the calculations that were performed in obtaining the Nth degree polynomial do not reduce the amount of work required to obtain polynomials of higher degree.

These difficulties can, however, be avoided by using an alternative representation of an approximating function. We write

$$g(x) = \sum_{m=0}^{M} a_m \varphi_m(x)\,, \tag{11.13}$$

where $\varphi_m(x)$ $(m=0,\ \ldots,\ M)$ is a function or polynomial. We show that, by proper choice of $\varphi_m(x)$, the least-squares approximation can be determined without having to solve a system of linear equations. Now the values of $a_m(x)$ $(m=0,\ \ldots,\ M)$ are chosen to minimize

$$e(a_0,\ldots,a_M) = \int_a^b \rho(x)(f(x) - g(x))^2\,dx\,,$$

where $\rho(x)$ is called a weight function. It is a nonnegative function, integrable on $[a,b]$, and it is also not identically zero on any subintervals of $[a,b]$. Proceeding as before gives

$$\sum_{m=0}^{M} a_m \int_a^b \rho(x)\varphi_m(x)\varphi_n(x)\,dx = \int_a^b \rho(x) f(x)\varphi_n(x)\,dx\,, \tag{11.14}$$

$$n=0,\ \ldots\ ,\ M.$$

This system will be particularly easy to solve if the functions $\varphi_m(x)$ are chosen to satisfy

$$\int_a^b \rho(x)\varphi_n(x)\varphi_m(x)\,dx = \begin{cases} 0 & \text{if } n \neq m \\ \alpha_n > 0 & \text{if } n = m \end{cases}. \tag{11.15}$$

In this case, all but one of the integrals on the left-hand side of equation (11.14) are zero, and solution can be obtained as

$$a_n = \frac{1}{\alpha_n}\int_a^b \rho(x) f(x)\varphi_n(x)\,dx\,.$$

Functions satisfying (11.15) are said to be orthogonal on the interval $[a,b]$ with respect to the weight function $\rho(x)$. Orthogonal function approximations have the advantage that an improvement of the approximation through addition of an extra term $a_{M+1}\varphi_{M+1}(x)$ does not affect the previously computed coefficients $a_0, \ldots, a_M$. Substitution of

$$x = \frac{b-a}{d-c}\overline{x} - \frac{cb-ad}{d-c}, \; dx = \frac{b-a}{d-c}d\overline{x}, \tag{11.16}$$

where $x \in [a,b]$ and $\overline{x} \in [c, d]$ in equations (11.13) to (11.15) yields a rescaled and/or shifted expansion interval.

Orthogonal functions (polynomials) play an important role in mathematical physics and mechanics, so approximation (11.13) may be very useful in various applications. For example, the Chebyshev polynomials can be applied to the problem of function approximation. The Chebyshev polynomials $T_n(x)$ are orthogonal on $[-1, 1]$ with respect to the weight function $\rho(x)=(1-x^2)^{-1/2}$. For $x \in [-1, 1]$ they are defined as $T_n(x)=\cos(n \cdot \arccos(x))$ for each $n \geq 0$ or

$$T_0(x) = 1, \; T_1(x) = x,$$

and

$$T_{n+1}(x) = 2xT_n(x) - T_{n-1}(x), \; n \geq 1.$$

The first few polynomials $T_n(x)$ are

$$\begin{aligned}
T_0(x) &= 1, \\
T_1(x) &= x, \\
T_2(x) &= 2x^2 - 1, \\
T_3(x) &= 4x^3 - 3x, \\
T_4(x) &= 8x^4 - 8x^2 + 1, \\
T_5(x) &= 16x^5 - 20x^3 + 5x, \\
T_6(x) &= 32x^6 - 48x^4 + 18x^2 - 1, \\
T_7(x) &= 64x^7 - 112x^5 + 56x^3 - 7x, \\
T_8(x) &= 128x^8 - 256x^6 + 160x^4 - 32x^2 + 1.
\end{aligned}$$

The orthogonality constant in (11.15) is $\alpha_n=\pi/2$. The Chebyshev polynomial $T_N(x)$ of degree $N \geq 1$ has N simple zeros in $[-1,1]$ at

$$x^*_{N,n} = \cos\left(\frac{2n-1}{2N}\pi\right) \tag{11.17}$$

for each $n=1, \ldots, N$. The error of polynomial approximation (11.13) is essentially equal to the omitted term $a_{M+1}T_{M+1}(x)$ for x in $[-1,1]$. Because $T_{M+1}(x)$ oscillates with amplitude one in the expansion interval, the maximum excursion of the absolute error in this interval is approximately equal to $|a_{M+1}|$.

Trigonometric functions are used to approximate functions with periodic behavior, that is, functions with the property $f(x+T)=f(x)$ for all x and $T>0$. Without loss of generality, we assume that $T=2\pi$ and restrict the approximation to the interval $[-\pi,\pi]$. Let us consider the following set of functions

$$\varphi_0(x) = \frac{1}{\sqrt{2\pi}},$$

$$\varphi_m(x) = \frac{1}{\sqrt{\pi}}\cos(mx),$$

$$m=1, \ldots, M,$$

$$\varphi_{M+m}(x) = \frac{1}{\sqrt{\pi}}\sin(mx),$$

$$m=1, \ldots, M-1.$$

Functions from this set are orthogonal on $[-\pi,\pi]$ with respect to the weight function $\rho(x)=1$, and α_k are equal to one. Given $f(x)$ is a continuous function on the interval $[-\pi,\pi]$, the least square approximation (11.16) is defined by

$$g(x) = \frac{a_0}{2\pi} + \frac{1}{\pi}\sum_{m=0}^{M} a_m \cos(mx) + \frac{1}{\pi}\sum_{m=1}^{M-1} b_m \sin(mx),$$

where

$$a_m = \int_{-\pi}^{\pi} f(x)\cos(mx)\,dx\,,\; m=0, \ldots, M,$$

$$b_m = \int_{-\pi}^{\pi} f(x)\sin(mx)\,dx\,,\; m=1, \ldots, M-1.$$

As $\cos(mx)$ and $\sin(mx)$ oscillate with an amplitude of one in the expansion interval, the maximum absolute error, that is $\max|f(x)-g(x)|$, may be estimated by

$$\frac{1}{\pi}(|a_M| + |b_{M-1}|).$$

The following function finds the coefficients for Chebyshev polynomial for a given input function argument and the degree of the polynomial.

Listing 11.10

```
function y = chebyshev(func, n)
x = zeros(1, n+1);
y = zeros(1, n+1);
k = 0:n;
z = (2*k+1)*pi/(2*n+2); x = cos(z);
f = feval(@exp, x);
T = zeros(n+1, n+1);
for i=1:n+1
    for j=1:n+1
```

```
        T(i, j) = cos((i-1)*z(j));
    end;
end;
for i=1:n+1;
    y(i) = sum(f.*T(i, :));
end;
y = y*2/(n+1);
y(1) = y(1)/2;
```

The following function evaluates the Chebyshev polynomial for a given input vector containing polynomial coefficients and the value of x where the polynomial needs to be calculated.

Listing 11.11

```
function y = chebypoly(c, x)
n = length(c); N = length(x);
for j=1:N
    y(j)= 0;
    for i=1:n
        y(j) = y(j) + c(i) * cos( (i-1) * acos(x(j)) );
    end;
end;
```

We use the last two function implementations in the following program to determine the Chebyshev approximation of the exponential function e^x for $x \in [-1, 1]$.

Listing 11.12

```
x = -1:0.1:1;
y = feval(@exp, x);
c = chebyshev(@exp, 5)
plot(x, y, 'x');
y2 = chebypoly(c, x);
hold on;
plot(x, y2);
hold off;
```

This program generates the following plot which shows some of the points of the exponential function along with the curve obtained by the approximated Chebyshev polynomial. It is interesting to see how the curve neatly goes through the exponential function points.

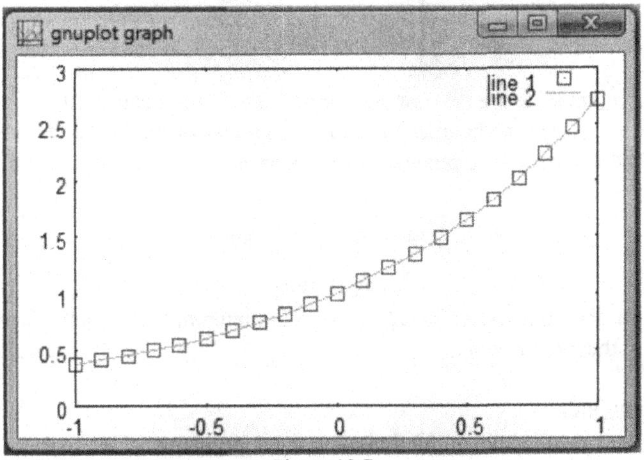

Figure 10.7

11.2.3 Approximation by interpolating polynomials

The interpolating polynomials may also be applied to the approximation of continuous functions $f(x)$ on some interval $[a,b]$. In this case $N+1$ distinct points x_0, ..., $x_N \in [a,b]$ are given, and $y_n=f(x_n)$. First, let us assume that $f(x)$ is defined on the interval $[-1, 1]$. If we construct an interpolating polynomial, which approximates function $f(x)$, then the error for this approximation can be represented in the form

$$f(x) - P_N(x) = \frac{1}{(N+1)!}\frac{d^{N+1}f}{dx^{N+1}}(z(x))\prod_{n=0}^{N}(x - x_n),$$

$$x\in[-1, 1],\ z(x)\in[-1, 1],$$

where $P_N(x)$ denotes the Lagrange interpolating polynomial. There is no control over $z(x)$, so minimization of the error is equivalent to minimization of the quantity $(x-x_0)(x-x_1)...(x-x_N)$. The way to do this is through the special placement of the nodes x_0, ..., x_N, since now we are free to choose these argument values. Before we turn to this question, some additional properties of Chebyshev polynomial should be considered.

The monic Chebyshev polynomial (a polynomial with leading coefficient equal to one) $T_n{}^*(x)$ is derived from the Chebyshev polynomial $T_n(x)$ by dividing its leading coefficients by 2^{n-1}, where $n\geq 1$. Thus

$$T_0^*(x) = 1,$$

$$T_n^*(x) = 2^{n-1}T_n(x), \text{ for each } n\geq 1.$$

As a result of the linear relationship between $T_n^*(x)$ and $T_n(x)$, the zeros of $T_n^*(x)$ are defined by expression (11.17). The polynomials $T_n^*(x)$, where $n\geq 1$, have a unique property: they have the least deviation from zero among all polynomials of degree n. Since $(x-x_0)(x-x_1)...(x-x_N)$ is a monic polynomial of degree $N+1$, the minimum of this quantity on the interval $[-1, 1]$ is obtained when

$$\prod_{n=0}^{N}(x - x_n) = T^*_{N+1}(x).$$

Therefore, x_n must be the $(n+1)$st zero of $T^*_{N+1}(x)$, for each n=0, …, N; that is, $x_n = x^*_{N+1,n+1}$. This technique for choosing points to minimize the interpolation error can be extended to a general closed interval $[a,b]$ by choosing the change of variable

$$\overline{x}_n = \frac{1}{2}\left[(b - a)x^*_{N+1,n+1} + (a + b)\right], \qquad (11.18)$$

$$\overline{x}_n \in [a,b],$$

to transform the numbers $x_n = x^*_{N+1,n+1}$ in the interval $[-1, 1]$ into corresponding numbers in the interval $[a,b]$.

11.3 Exercises

It is suggested to solve one of the following three problems:

(a) Calculate coefficients of Lagrange interpolating polynomial that passes through the given points x_n and y_n (tables 11.2–11.7). Plot the polynomial together with initial data.

(b) Calculate coefficients of trigonometric interpolating polynomial that passes through the given points x_n and y_n (tables 11.8–11.13). Plot the polynomial together with initial data.

(c) Calculate coefficients of the Lagrange interpolating polynomial that approximates given function on some interval. Plot the function and polynomial. Suggested functions and intervals are

(1) Bessel function $J_0(x), \quad x \in [0,4]$,

(2) Bessel function $J_1(x), \quad x \in [0,3]$,

(3) Error function $\mathrm{erf}(x) = \frac{2}{\sqrt{\pi}}\int_0^x \exp(-t^2)\,dt, \quad x \in [0,4]$,

(4) Function $E_1(x) = \int_x^\infty \frac{\exp(-t)}{t}\,dt, \quad x \in [0.1,\ 2]$,

(5) Inverse hyperbolic sinus $\mathrm{arcsh}(x), \quad x \in [-5,5]$,

(6) Inverse hyperbolic cosine $\mathrm{arcth}(x), \quad x \in [-0.9,0.9]$.

Methodical comments: to compute the values of suggested functions use intrinsic functions besselj$(n,\ x)$, erf(x), expint(x), asinh(x), atanh(x).

Table 11.2

x	Y
0	0
0.2	0.3696
0.5	0.5348
1	0.6931
1.5	0.7996
2	0.8814

Table 11.3

x	y
0	0
0.2	0.1753
0.5	0.3244
1	0.3466
1.5	0.2819
2	0.2197

Table 11.4

x	y
0	1
0.2	0.8004
0.5	0.5134
1	0.1460
1.5	0.0020
2	0.0577

Table 11.5

x	y
0	0
0.2	0.0329
0.5	0.1532
1	0.3540
1.5	0.3980
2	0.2756

Table 11.6

x	y
0	0
0.2	0.0079
0.5	0.1149
1	0.7081
1.5	1.4925
2	1.6536

Table 11.7

x	y
0	0
0.2	0.6687
0.5	0.8409
1	1
1.5	1.1067
2	1.1892

Table 11.8

x	y
0	0
0,3π	0.2164
0,5π	0.3163
0,8π	0.3585
π	0.3173
1,2π	0.2393
1,6π	0.0686
1,8π	0.0169

Table 11.9

x	y
0	0
0,3π	0.2295
0,5π	0.2018
0,8π	0.1430
π	0.1012
1,2π	0.0636
1,6π	0.0137
1,8π	0.0030

Table 11.10

x	y
0	0
0,3π	0.1769
0,5π	0.1470
0,8π	0.0770
π	0.0432
1,2π	0.0219
1,6π	0.0039
1,8π	0.0011

Table 11.11

x	y
0	1.0
0,3π	0.7939
0,5π	0.5
0,8π	0.0955
π	0
1,2π	0.0955
1,6π	0.6545
1,8π	0.9045

Table 11.12

x	y
0	0
0,3π	0.1943
0,5π	0.7854
0,8π	2.2733
π	3.1416
1,2π	3.4099
1,6π	1.7366
1,8π	0.5400

Table 11.13

x	y
0	0
0,3π	0.2001
0,5π	0.6267
0,8π	1.4339
π	1.7725
1,2π	1.7562
1,6π	0.7746
1,8π	0.2271

Bibliography

Ascher U. M. and Petzold L. R., Computer methods for ordinary differential equations and differential-algebraic equations, SIAM, Philadelphia, 1998.

Butcher J. C., The numerical analysis of ordinary differential equations: Runge-Kutta and general linear methods, John Wiley & Sons, 1987.

Chonacky, N., 3Ms For Instruction, Computing in Science & Engineering, May/June 2005.

Chonacky, N., 3Ms For Instruction Part 2, Computing in Science & Engineering, July/August 2005.

Dormand J. R., Numerical methods for differential equations, CRC Press, 1996.

Eaton, J., Ten Years of Octave – Recent Developments and Plans for the Future, Technical report number 2003-01, Texas-Wisconsin Modeling and Control Consortium.

Eclipse website: http://www.eclipse.org.

Faddeev D. K. and Faddeeva V. N., Computational methods of linear algebra, W. H. Freeman, San Francisco, 1963.

GnuPlot project website: http://www.gnuplot.info.

Godunov S. K. and Ryabenkii V. S., Differential schemes: an introduction to the underlying theory, Elsevier Science, 1987.

Greenbaum A., Iterative methods for solving linear systems, SIAM, Philadelphia, 1997.

Hageman L. A. and Young D. M., Applied iterative methods, Academic Press, 1981.

Hairer E., Nørsett S. P. and Wanner G., Solving ordinary differential equations I., Springer-Verlag, 1987.

Hairer E. and Wanner G., Solving ordinary differential equations II., Springer-Verlag, 1991.

Hall G. and Watt J. M. (Eds), Modern numerical methods for ordinary differential equations, Clarendon Press, Oxford, 1976.

Hudak D. E, Developing a Computational Science IDE for HPC Systems, Third International Workshop on Software Engineering for High Performance Computing Applications, 2007.

Johnson, G., LabVIEW Graphical Programming: Practical Applications in Instrumentation and Control, McGraw-Hill, ISBN: 007032915X.

Kelly C. T., Iterative methods for linear and nonlinear equations, SIAM, Philadelphia, 1995.

Kiusalaas, J., Numerical Methods in Engineering with Python, Cambridge University Press, 2005.

Lambert J. D., Computational methods in ordinary differential equations, John Wiley & Sons, 1973.

Lambert J. D., Numerical methods for ordinary differential systems: the initial value problem, John Wiley & Sons, 1991.

Luszczek P., Design of Interactive Environment for Numerically Intensive Parallel Linear Algebra Calculations, Lecture Notes in Computer Science 3039, Springer-Verlag Berlin-Heidelberg, March 2004.

Matlab website: http://www.mathworks.com.

numEclipse project: http://www.numEclipse.org.

Parlett B. N., The symmetric eigenvalue problem, Prentice Hall, 1980.

Pires, P., Free/open source software: An alternative for engineering students, 32nd ASEE/IEEE Frontiers in Education Conference, November 6–9, 2002, Boston, MA.

PLPlot website: http://plplot.sourceforge.net.

Späth H., One-dimensional spline interpolation algorithms, A. K. Peters, Massachusetts, 1995.

Spitaleri, R., A scientific computing environment for differential field simulation, Mathematics and Computers in Simulation 63 (2003) 79–91.

Traub J. F., Iterative methods for the solution of equations, Prentice Hall, 1964.

Young D. M., Iterative solution of large linear systems, Academic Press, 1971.

Printed in the United States
By Bookmasters